Lecture Notes in Computer Science　　　11223

Commenced Publication in 1973
Founding and Former Series Editors:
Gerhard Goos, Juris Hartmanis, and Jan van Leeuwen

Editorial Board

More information about this series at http://www.springer.com/series/7409

Stéphane Marchand-Maillet · Yasin N. Silva
Edgar Chávez (Eds.)

Similarity Search and Applications

11th International Conference, SISAP 2018
Lima, Peru, October 7–9, 2018
Proceedings

 Springer

Editors
Stéphane Marchand-Maillet
University of Geneva
Carouge
Switzerland

Yasin N. Silva
Arizona State University
Tempe, AZ
USA

Edgar Chávez
Center for Scientific Research and Higher
  Education
Ensenada
Mexico

ISSN 0302-9743   ISSN 1611-3349 (electronic)
Lecture Notes in Computer Science
ISBN 978-3-030-02223-5   ISBN 978-3-030-02224-2 (eBook)
https://doi.org/10.1007/978-3-030-02224-2

Library of Congress Control Number: 2018957279

LNCS Sublibrary: SL3 – Information Systems and Applications, incl. Internet/Web, and HCI

This Springer imprint is published by the registered company Springer Nature Switzerland AG
The registered company address is: Gewerbestrasse 11, 6330 Cham, Switzerland

Preface

This volume contains the papers presented at the 11th International Conference on Similarity Search and Applications (SISAP 2018) held in Lima, Peru, during October 7–9, 2018.

SISAP is an annual forum for researchers and application developers in the area of similarity data management. It focuses on the technological problems shared by numerous application domains, such as data mining, information retrieval, multimedia, computer vision, pattern recognition, computational biology, geography, biometrics, machine learning, and many others that make use of similarity search as a necessary supporting service.

From its roots as a regional workshop in metric indexing, SISAP has expanded to become the only international conference entirely devoted to the issues surrounding the theory, design, analysis, practice, and application of content-based and feature-based similarity search. The SISAP initiative has also created a repository (http://www.sisap. org/) serving the similarity search community, for the exchange of examples of real-world applications, source code for similarity indexes, and experimental test beds and benchmark data sets.

The call for papers welcomed full papers, short papers, as well as demonstration papers, with all manuscripts presenting previously unpublished research contributions.

We received 31 submissions from authors based in 17 different countries. The Program Committee (PC) was composed of 50 international members. Reviews were thoroughly discussed by the chairs and PC members: Each submission received three reviews. Based on these reviews and discussions among PC members, the PC chairs accepted 16 full papers, three short papers, and one demonstration paper to be included in the conference program and the proceedings. At SISAP 2018, all contributions were presented orally.

The proceedings of SISAP are published by Springer as a volume in the *Lecture Notes in Computer Science* (LNCS) series. For SISAP 2018, as in previous years, extended versions of five selected excellent papers were invited for publication in a special issue of the journal *Information Systems*. The conference also conferred a Best Paper Award, as judged by the PC co-chairs and Steering Committee.

Beside the presentations of the accepted papers, the conference program featured three keynote presentations from exceptionally skilled scientists: Prof. Alistair Moffat from the University of Melbourne, Australia, Prof. Hanan Samet from the University of Maryland, USA, and Prof. Moshe Y. Vardi from Rice University, USA.

We would like to thank all the authors who submitted papers to SISAP 2018. We would also like to thank all members of the PC and the external reviewers for their effort and contribution to the conference. We want to express our gratitude to the members of the Organizing Committee for the enormous amount of work they did.

We also thank our sponsors and supporters for their generosity. All the submission, reviewing, and proceedings generation processes were made much easier through the EasyChair platform.

August 2018   Stéphane Marchand-Maillet
Yasin N. Silva
Edgar Chávez

Organization

General Chair

Edgar Chavez   CICESE, Mexico

Program Chairs

Stéphane Marchand-Maillet   Viper Group - University of Geneva, Switzerland
Yasin N. Silva   Arizona State University, USA

Program Committee

Giuseppe Amato	ISTI-CNR, Italy
Laurent Amsaleg	CNRS-IRISA, France
Panagiotis Bouros	Aarhus University, Denmark
Nieves R. Brisaboa	Universidade da Coruña, Spain
Benjamin Bustos	University of Chile, Chile
K. Selcuk Candan	Arizona State University, USA
Aniket Chakrabarti	Microsoft, USA
Edgar Chavez	CICESE, Mexico
Paolo Ciaccia	University of Bologna, Italy
Richard Connor	University of Strathclyde, UK
Michel Crucianu	CNAM, France
Vlad Estivill-Castro	Griffith University, Australia
Fabrizio Falchi	ISTI-CNR, Italy
Karina Figueroa	Universidad Michoacana, Mexico
Teddy Furon	Inria, France
Claudio Gennaro	ISTI-CNR, Italy
Costantino Grana	University of Modena and Reggio Emilia, Italy
Michael E. Houle	National Institute of Informatics, Japan
Ichiro Ide	Nagoya University, Japan
Yoshiharu Ishikawa	Nagoya University, Japan
Jakub Lokoc	Charles University in Prague, Czech Republic
Luisa Mico	University of Alicante, Spain
Henning Müller	HES-SO, Switzerland
Vo Ngoc Phu	Institute of Research and Development, Duy Tan University, Da Nang, Vietnam
Vincent Oria	NJIT, USA
Deepak P.	Queen's University Belfast, UK
Apostolos N. Papadopoulos	Aristotle University of Thessaloniki, Greece
Rodrigo Paredes	Universidad de Talca, Chile
Marco Patella	University of Bologna, Italy

Contents

Clustering and Outlier Detection

Graphs and Applications

Shared Session SISAP and SPIRE

Metric Search

Re-ranking Permutation-Based Candidate Sets with the n-Simplex Projection

Giuseppe Amato[1], Edgar Chávez[2], Richard Connor[3], Fabrizio Falchi[1],
Claudio Gennaro[1], and Lucia Vadicamo[1(✉)]

[1] Institute of Information Science and Technologies (ISTI), CNR, Pisa, Italy
{giuseppe.amato,fabrizio.falchi,claudio.gennaro,
lucia.vadicamo}@isti.cnr.it
[2] Department of Computer Science, CICESE, Ensenada, Mexico
elchavez@cicese.mx
[3] Department of Computing Science, University of Stirling,
Stirling FK9 4LA, Scotland
richard.connor@stir.ac.uk

Abstract. In the realm of metric search, the permutation-based approaches have shown very good performance in indexing and supporting approximate search on large databases. These methods embed the metric objects into a permutation space where candidate results to a given query can be efficiently identified. Typically, to achieve high effectiveness, the permutation-based result set is refined by directly comparing each candidate object to the query one. Therefore, one drawback of these approaches is that the original dataset needs to be stored and then accessed during the refining step. We propose a refining approach based on a metric embedding, called n-Simplex projection, that can be used on metric spaces meeting the n-point property. The n-Simplex projection provides upper- and lower-bounds of the actual distance, derived using the distances between the data objects and a finite set of pivots. We propose to reuse the distances computed for building the data permutations to derive these bounds and we show how to use them to improve the permutation-based results. Our approach is particularly advantageous for all the cases in which the traditional refining step is too costly, e.g. very large dataset or very expensive metric function.

Keywords: Metric search · Permutation-based indexing
n-point property · n-Simplex projection · Metric embedding
Distance bounds

1 Introduction

The problem of searching data objects that are close to a given query object, under some metric function, has a vast number of applications in many branches of computer science, including pattern recognition, computational biology and multimedia information retrieval, to name but a few. This search paradigm,

© Springer Nature Switzerland AG 2018
S. Marchand-Maillet et al. (Eds.): SISAP 2018, LNCS 11223, pp. 3–17, 2018.
https://doi.org/10.1007/978-3-030-02224-2_1

referred to as *metric search*, is based on the assumption that data objects are represented as elements of a metric space (D, d) where the *metric*[1] function $d : D \times D \rightarrow \mathbb{R}^+$ provides a measure of the closeness of the data objects.

In metric search, the main concern is processing and structuring a finite set of data $X \subset D$ so that *proximity queries* can be answered quickly and with a low computational cost. A proximity query is defined by a query object $q \in D$ and a proximity condition, such as "find all the objects within a threshold distance of q" (*range query*) or "finding the k closest objects to q" (*k-nearest neighbour query*). The response to a query is the set of all the objects $o \in X$ that satisfy the considered proximity condition. Providing an exact response is not feasible if the search space is very large or if it has a high intrinsic dimensionality since a large fraction of the data needs to be inspected to process the query. In such cases, the exact search rarely outperforms a sequential scan [22]. To overcome the *curse of dimensionality* [19] researchers proposed several *approximate search* methods that are less (but still) affected by this phenomenon.

Many approximate methods are based on the idea of mapping the data objects into a more tractable space in which we can efficiently perform the search. Successful examples are the *Permutation-Based Indexing* (PBI) approaches that represent data objects as a sequence of identifiers (*permutation*). Typically, the permutation for an object o is computed as a ranking list of some preselected reference points (*pivots*) according to their distance to o. The main rationale behind this approach is that if two objects are very close one to the other, they will sort the set of pivots in a very similar way, and thus the corresponding permutation representations will be close as well. The search in the permutation space is used to build a candidate result set that is normally refined by comparing each candidate object to the query one (according to the metric governing the data space). This refinement step therefore requires access to the original data, which is likely to be too large to fit into main memory. However, some kind of refinement step is likely to be required as the search in the permutation space typically has relatively low precision.

In this paper, we focus on the k-nearest neighbour (k-NN) query search and we investigate several approaches to perform the refining step without accessing the original data, but instead exploiting the distances between the objects and the pivots (calculated at indexing time and stored within the permutations) and the distances between the query and the pivots (evaluated when computing the query permutation). In particular, for a large class of metric spaces that meet the so-called "*n-point property*" [9,11] we propose the use of the *n-Simplex projection* [12] that allows mapping metric objects into a finite dimensional Euclidean space where upper- and lower-bounds for the original distances can be calculated. We show how these distance bounds can be used to refine the permutation-based results, therefore avoiding access to the original dataset.

---

[1] Throughout this paper, we use the term "metric" and "distance" interchangeably to indicate a function satisfying the metric postulates [23].

2   Related Work

The idea of approximating the distance between any two metric objects by comparing their permutation-based representations was originally proposed in [5,8]. Several techniques for indexing and searching permutations were proposed in literature, including indexes based on inverted files, like the *Metric Inverted File* (MI-File) [4] and its variants, or using prefix trees, like the *Permutation Prefix Index* (PP-Index) [13] and the *Pivot Permutation Prefix Index* (PPP-Index) [17]. The permutation-based approach are *filter* and *refine* methods: a candidate result set is identified by performing the search in the permutation space, then the result set is refined, commonly, by evaluating the actual distance between the query and the candidate objects.

The permutation representation of an object is computed by ordering the identifiers of a set of pivots according to their distances to the object [3]. However, the computation of these distances is just one, yet effective, approach to associate a permutation to each data object. For example, the *Deep Permutations* [2] have been recently proposed as an efficient and effective alternative for generating permutations of emerging deep features. However, this approach is suitable only for specific data domains while the traditional approach is generally applicable since it requires only the existence of a distance function to compare data objects.

The distances between the data objects and a set of pivots can be used also to embed the data into another metric space where it is possible to deduce upper- and lower-bounds on the actual distance of any pair of objects. In this context, one of the very first embeddings proposed in a metric search scenario was the one representing each data object with a vector of its distances to the pivots. The LAESA [16] is a notable example of indexing technique using this approach. Recently, Connor et al. [10–12] observed that for a large class of metric spaces it is possible to use the distances to a set of n pivots to project the data objects into a n-dimensional Euclidean space such that in the projected space (1) the distances object-pivots are preserved, (2) the Euclidean distance between any two points is a lower-bound of the actual distance, (3) also an upper-bound can be easily computed. They called this approach *n-Simplex projection* and they proved that it can be used in all the metric spaces meeting the *n-point property* [7]. As also pointed out in [9], many common metric spaces meet the desired property, like Cartesian spaces of any dimension with the Euclidean, cosine or quadratic form distances, probability spaces with the Jenson-Shannon or the Triangular distance, and more generally any Hilbert-embeddable space [7,20].

3   Background

In the following, we summarize key concepts of some metric space transformations based on the use of distances between data objects and a set of pivots. The rationale behind these approaches is to project the original data into a space that has better indexing properties than the original, or where the comparison between objects is less expensive than the original distance. In particular, we review data embeddings into permutation spaces, where objects can be efficiently indexed using PBI methods, and other pivot-based embeddings that allow computing upper- and lower-bounds of the actual distance. Table 1 summarizes the notation used.

Table 1. Notation used throughout this paper

Symbol	Definition
(D, d)	Metric space
X	Finite search space, $X \subseteq D$
$\{p_1, \ldots, p_n\}$	Set of pivots, $p_i \in D$
n	Number of pivots
o, s	Data objects, $o, s \in X$
q	Query, $q \in D$
k, k'	Number of results of a nearest neighbour search
amp	Amplification factor
Π_o	Pivot permutation
Π_o^{-1}	Inverted permutation
l	Location parameter (permutation prefix length)
$\Pi_{o,l}$	Truncated permutation (permutation prefix of length l)
$\Pi_{o,l}^{-1}$	Inverted truncated permutation
$PivotSet(\Pi_{o,l})$	The pivots whose identifiers appear in $\Pi_{o,l}$
$\Gamma_{o,q}$	Pivots in the intersection $PivotSet(\Pi_{q,l}) \cap PivotSet(\Pi_{o,l})$
$S_{\rho,l}$	Spearman's rho with location parameter l
ℓ_2	Euclidean distance
ℓ_∞	Chebyshev distance
$\lvert \cdot \rvert$	Size of a set

3.1   Permutation-Based Representation

Let $\mathcal{D}$ a data domain and $d : \mathcal{D} \times \mathcal{D} \rightarrow \mathbb{R}^+$ a *metric* function on it[2]. A permutation-based representation Π_o (briefly *permutation*) of an object $o \in \mathcal{D}$ with respect to a fixed set of *pivots*, $\{p_1, \ldots, p_n\} \subset \mathcal{D}$, is the sequence of pivots identifiers ordered by their distance to o.

Formally, the permutation $\Pi_o = [\Pi_o(1), \Pi_o(2), \ldots, \Pi_o(n)]$ lists the pivot identifiers $\{1, \ldots, n\}$ in an order such that $\forall\, i, j \in \{1, \ldots, n\}$

$$\Pi_o(i) < \Pi_o(j) \Leftrightarrow \begin{matrix} d(o, p_{\Pi_o(i)}) < d(o, p_{\Pi_o(j)}) \\ or \\ \left(d(o, p_{\Pi_o(i)}) = d(o, p_{\Pi_o(j)}) \wedge (i < j) \right) \end{matrix} \qquad (1)$$

---

[2] In this work, we focus on metric search. The requirement that the function d satisfies the metric postulates is sufficient, but not necessary, to produce a permutation-based representation. For example, d may be a dissimilarity function.

An equivalent permutation-based representation is the *inverted permutation*, defined as $\Pi_o^{-1} = [\Pi_o^{-1}(1), \Pi_o^{-1}(2), \ldots, \Pi_o^{-1}(n)]$, where $\Pi_o^{-1}(i)$ denotes the position of a pivot p_i in the permutation Π_o. The inverted permutation is such that $\Pi_o(\Pi_o^{-1}(i)) = i$. Note that the value at the coordinate i in the permutation Π_o is the identifier of the pivot at i-th position in the ranked list of the nearest pivots to o; the value at the coordinate i in the inverted representation Π_o^{-1} is the rank of the pivot p_i in the list of the nearest pivots to o.

The inverted permutation representation is often used in practice since it allows us to represent permutations in a Cartesian coordinate system and easily compute most of the commonly-used distances between permutations as distances between Cartesian points. In this paper, we use the Spearman Rho that is defined as $S_\rho(\Pi_o, \Pi_s) = \ell_2(\Pi_o^{-1}, \Pi_s^{-1})$ for any two permutations Π_o, Π_s.

Most of the PBI methods, e.g. [4,13,17], use only a fixed-length prefix of the permutations in order to represent or compare objects. This choice is based on the intuition that the most relevant information in the permutation is present in its very first elements, i.e. the identifiers of the closest pivots. Moreover, using the positions of the nearest l out of n pivots often leads to obtaining better or similar effectiveness to using the full-length permutation [3,4], resulting also in a more compact data encoding. The permutation prefixes are compared using *top-l distances* [14], like the Spearman Rho with location parameter l defined as $S_{\rho,l}(\Pi_o, \Pi_s) = \ell_2(\Pi_{o,l}^{-1}, \Pi_{s,l}^{-1})$, where $\Pi_{o,l}^{-1}$ is the *inverted truncated permutation*:

$$\Pi_{o,l}^{-1}(i) = \begin{cases} \Pi_o^{-1}(i) & \text{if } \Pi_o^{-1}(i) \leq l \\ l+1 & \text{otherwise} \end{cases} \qquad (2)$$

3.2   Pivoted Embedding

The distances between metric objects and a set of pivots $\{p_1, \ldots, p_n\} \subset \mathcal{D}$ can be also used to embed a metric space into $(\mathbb{R}^n, \ell_\infty)$:

$$f_n : (D, d) \rightarrow (\mathbb{R}^n, \ell_\infty)$$
$$o \rightarrow [d(o, p_1), \ldots, d(o, p_n)]$$

Using the triangle inequality of the metric governing the space is possible to prove that

$$\max_{i=1,\ldots,n} |d(o, p_i) - d(s, p_i)| \leq d(o, s) \leq \min_{i=1,\ldots,n} |d(o, p_i) + d(s, p_i)| \qquad (3)$$

which means that $\ell_\infty(f_n(o), f_n(s))$ is a lower-bound of $d(o, s)$ and that also an upper-bound can be defined using the projected objects $f_n(o), f_n(s)$ (see [23, p. 28]). In the following we referred to these bounds to as *Pivoted embedding* bounds. Please note that if we use just a subset of size l of the pivots $\{p_1, \ldots, p_n\}$, the corresponding mapping f_l provides upper- and lower-bounds that are less tight than that obtained using f_n.

This family of embeddings are typically used in indexing tables like LAESA [16] or for space pruning [23]. However, as further described in Sect. 4, in this work we used them not for indexing purpose, but rather as techniques to approximate the distances between a query and data objects already indexed using a permutation-based approach.

3.3 n-Simplex Projection

In [9,12] it was observed that there exists a large class of metric spaces that satisfy the so-called *n-point property*, which provides geometric guarantees stronger than triangle inequality.

A metric space meets the *n*-point property if, and only if, any set of *n* points of the space can be embedded into a $(n-1)$-dimensional Euclidean space while preserving all the $\binom{n}{2}$ inter-points distances. This property was exploited in [12] to define an embedding of the considered metric space into a finite-dimensional Euclidean space. Specifically, they defined a family of functions $\phi_n : (D, d) \rightarrow (\mathbb{R}^n, \ell_2)$, where $\phi_n(o)$ is obtained using the distances between o and a set of pivots $\{p_1, \ldots, p_n\} \subset \mathcal{D}$. They provided also an inductive algorithm for determining the Cartesian coordinates of $\phi_n(o)$ that, given the distances $d(o, p_i)$, requires the computations of $O(n)$ Euclidean distances between vectors having less than n dimensions. The core idea of their approach is computing the vector $\phi_n(o)$ as the apex of a *n*-dimensional simplex[3] where the length of the *i*-th edge connecting the apex and a simplex base corresponds to the actual distance $d(o, p_i)$. The simplex base is computed using the distances $d(p_i, p_j)$ for all $i, j \in \{1, \ldots, n\}$.

One of the main outcomes of this embedding is that it allows deriving upper- and lower-bounds of the actual distance by computing the Euclidean distance between two Cartesian points. In facts, given the apexes

$$\phi_n(o) = [x_1, x_2, \ldots, x_{n-1}, x_n]$$
$$\phi_n(s) = [y_1, y_2, \ldots, y_{n-1}, y_n]$$

it holds

$$\sqrt{\sum_{i=1}^{n}(x_i - y_i)^2} \leq d(o, s) \leq \sqrt{\sum_{i=1}^{n-1}(x_i - y_i)^2 + (x_n + y_n)^2} \tag{4}$$

So, if defining $\phi_n^-(s) = [y_1, y_2, \ldots, y_{n-1}, -y_n]$, we have that $\ell_2(\phi_n(o), \phi_n(s))$ and $\ell_2(\phi_n(o), \phi_n^-(s))$ are respectively a lower- and upper-bound for $d(o, s)$. Connor et al. [12][4] proved that, if ϕ_n is the *n*-Simplex projection based on the pivots $\{p_1, \ldots, p_n\}$, and ϕ_m is the *m*-Simplex projection based on the pivots $\{p_1, \ldots, p_n, p_{n+1}, \ldots, p_m\}$ then

$$\ell_2(\phi_n(o), \phi_n(s)) \leq \ell_2(\phi_m(o), \phi_m(s)) \leq d(o, s) \leq \ell_2(\phi_m(o), \phi_m^-(s)) \leq \ell_2(\phi_n(o), \phi_n^-(s)).$$

Moreover, they experimentally showed that the so-defined distance bounds converge to the actual distance when increasing the number of pivots.

---

[3] A simplex is a generalisation of a triangle or a tetrahedron in arbitrary dimensions. We refer to [12] for further details.

[4] See also the on-line Appendix at http://arxiv.org/abs/1707.08370.

4   Re-ranking Permutation-Based Candidate Set

The permutation-based methods are filter-and-refine approaches that map original data (X, d) into a permutation space. The permutation representations are used to identify a set of candidate results for a given query $q \in D$. The candidate results are then refined, typically by comparing the candidate objects with the query one according to the actual distance d. In the following, we investigate the use of other refining approaches to answer a k-NN query. The aim is improving the permutation-based results while getting rid of the original dataset.

Let $CandSet(q)$ the set of candidate results selected using the permutation-based encoding, where $|CandSet(q)| = k' \geq k$. The candidate result set can be built, for example, by performing a k'-NN search in the permutation space (e.g. using the MI-File [4]) or by finding objects with a common permutation prefix (e.g. using the PP-codes [13]). In any case here we assume to have access only to the permutation prefixes and not to the full-length permutations, as done in many PBI approach [4,13,17].

Let $PivotSet(\Pi_{o,l})$ the set of the l closest pivots to the object o, i.e. the pivots whose identifiers appear in the prefix permutation $\Pi_{o,l}$. We assume that the distances between each object and its l closest pivots are stored and indexed within the object prefix permutation. This can be done with a slight modification of the used permutation-based index. In the following, we assume that the objects are indexed using inverted files. Figure 1 shows a naive example for integrating the object-pivot distances into the posting lists, such as the ones used in the MI-file [4]. However, the approach presented in this paper can be extended to cope with different permutation-based indexes.

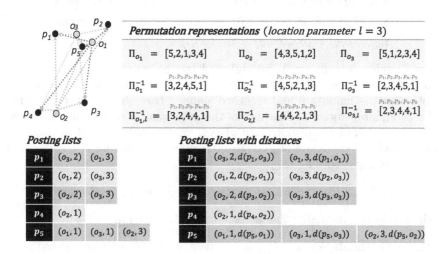

Fig. 1. Example of posting lists and posting list with distances generated to index three objects using five pivots and a location parameter $l = 3$

We propose to refine the candidate result set by selecting the top-k candidate objects ranked according to some pivot-based dissimilarity function. To this end, we tested the *Pivoted embedding* and the *n-Simplex projection* distance bounds, computed using the metric mapping described in Sects. 3.2 and 3.3. Specifically, at query time, for each object $o \in CandSet(q)$ we approximate the actual distance $d(o, q)$ on the basis of the distances $d(q, p_j)$, $d(o, p_j)$ for $p_j \in \Gamma_{o,q} = PivotSet(\Pi_{q,l}) \cap PivotSet(\Pi_{o,l})$ as follows:

Pivoted embedding - As a consequence of Eq. 3 we have

$$\max_{p_j \in \Gamma_{o,q}} |d(o, p_j) - d(q, p_j)| \leq d(o, q) \leq \min_{p_j \in \Gamma_{o,q}} |d(o, p_i) + d(q, p_i)| \qquad (5)$$

so we consider three possible re-rankings of the candidate objects, based on the following dissimilarity measures

$$P_{lwb}(o, q) = \max_{p_j \in \Gamma_{o,q}} |d(o, p_j) - d(q, p_j)| \qquad \text{lower-bound}$$

$$P_{upb}(o, q) = \min_{p_j \in \Gamma_{o,q}} (d(o, p_j) + d(q, p_j)) \qquad \text{upper-bound}$$

$$P_{mean}(o, q) = (P_{upb}(o, q) + P_{lwb}(o, q))/2 \qquad \text{mean}$$

Simplex projection - For each candidate object o, the pivots in $\Gamma_{o,q}$ are used to build a simplex base. The simplex base and the distances $d(o, p_j)$, $d(q, p_j)$ with $p_j \in \Gamma_{o,q}$ are used to compute the apexes $\phi_h(o), \phi_h(q), \phi_h^-(q) \in \mathbb{R}^h$, where $h = |\Gamma_{o,q}| \leq l$. We consider the re-rankings of the candidate objects based on the following dissimilarity measures:

$$S_{lwb}(o, q) = \ell_2(\phi_h(o), \phi_h(q)) \qquad \text{lower-bound}$$

$$S_{upb}(o, q) = \ell_2(\phi_h(o), \phi_h^-(q)) \qquad \text{upper-bound}$$

$$S_{mean}(o, q) = (S_{upb}(o, q) + S_{lwb}(o, q))/2 \qquad \text{mean}$$

The Simplex bounds are highly affected by the number h of pivots used to build the simplex base (the higher h, the tighter the bounds), moreover note that the number h and the used simplex base change when changing the candidate object o. This means that the quality of the simplex-based approximation of the distance $d(o, q)$ may vary significantly when changing the considered candidate object. To overcome this issue, we also considered the re-ranking according to

$$SN_{mean}(o, q) = S_{mean}(o, q)/g(h) \qquad \text{normalized mean}$$

where $g(h)$ is a normalization factor, further discussed in Sect. 5.2.

The lower-bounds S_{lwb} and P_{lwb} are metrics, while the other considered measures are just dissimilarity functions.

Note that for all those approaches no new object-pivot distances are evaluated at either indexing or query time, since the used distances are already computed for building the permutation-based representations of the objects/query.

Moreover, the distances $d(o, p_j)$ with $p_j \in \Gamma_{o,q}$ are retrieved while scanning the posting list to build the candidate result set, therefore the considered re-ranking approaches do not require further disk accesses in addition to the index accesses already made to find the candidate results.

5 Experiments

In this section, we evaluate the quality of the re-ranking approach discussed above. We first describe the experimental setup and then we report results and their analysis.

5.1 Experimental Settings

The experiments were conducted using three publicly available datasets, namely YFCC100M [21], Twitter-Glove [18], and SISAP colors [15].

YFCC100M collection contains almost 100M images, all uploaded to Flickr between 2004 and 2014. In the experiments, we used a subset of 1M deep Convolutional Neural Network features extracted by Amato et al. [1] and available at http://www.deepfeatures.org/. Specifically, we used the activations of the *fc6* layer of the HybridNet [24] after ReLu and ℓ_2 normalization. The resulting features are 4,096-dimensional vectors. We followed the common choice of using the *Euclidean distance* to compare them.

Twitter-GloVe is a collection of 1.2M GloVe [18] features (word embeddings) trained on tweets. We used the 100-dimensional pre-trained word vector available at https://nlp.stanford.edu/projects/glove/. These word vectors are often used as vocabulary terms to embed a document into a vector representation, for example by averaging the vectors of the terms contained in the text. In such cases, the space of the vocabulary word embeddings is representative of the space of the document embeddings. We used the *Cosine distance*, i.e. $d_{\mathrm{Cos}}(x, y) = \sqrt{1 - \frac{x \cdot y}{\|x\|_2 \|y\|_2}}$, to compare the GloVe vectors.

SISAP colors is a commonly used benchmark for metric indexing approaches. It contains about 113 K feature vectors of dimensions 112, representing color histograms of medical images. We used the *Jenson-Shannon distance*, defined as in [9], for the feature comparison.

For each dataset we build a ground-truth for exact similarity search related to 1,000 randomly-selected queries. The ground-truths are used to evaluate the quality of the approximate results obtained by re-ranking a permutation-based result set of size $k' \geq k$. Specifically, we select as a candidate result set for a k-NN query the set obtained by performing a k'-NN search in the permutation space. Then we re-rank the candidate results and we select the top-k objects. The quality of the so obtained approximate results was evaluated using the *recall@k*, defined as $|\mathcal{R} \cap \mathcal{R}^A|/k$ where $\mathcal{R}$ is the result set of the exact k-NN search in the original metric space and $\mathcal{R}^A$ is the approximate result set. We set $k = 10$ and $k' = 100$, thus, in order to build the candidate result set, we performed a 100-NN search in the permutation space using the Spearman's rho with location parameter l. In all the tests the pivots were randomly selected. We used about $4,000$ pivots for YFCC100M and Twitter-GloVe, and $n = 1,000$ pivots for the smaller SISAP colors dataset.

5.2 Results

Figure 2 reports the $recall@10$ with respect to the length l of the permutation prefixes used to represent the data objects. Please note that, in each test, we fixed the number n of pivots and we varied the prefix length l.

In order to evaluate the permutation-based results without any re-ranking, we selected the first $k = 10$ objects of the candidate set, ordered according to their permutation-based distance to the query that, in our case, was $S_{\rho,l}$. This baseline approach is indicated as no reordering in the graphs. We compared it with the re-rankings based on the Pivoted embedding and the Simplex projection distance approximations (lower-bound, upper-bound and mean). In all cases, it is important to keep in mind that, for a fixed value l and for a candidate object o, the number h of pivots used to compute the distance approximations is less than l; moreover, it varies when changing the candidate object because it equals the cardinality of $\Gamma_{o,q}$. So, typically h is greater for objects in top positions in the permutation-based candidate set and decrease for far objects. Moreover, the greater the l, the greater the h and so the better the approximation bounds.

Surprisingly, we observed that in almost all the tested cases the Pivoted embedding approach greatly degrades the quality of the permutation-based results. Moreover, on YFCC100M and Twitter-GloVe sets it never reaches a $recall$ greater than 0.3. So, in our tests, the Pivoted distance approximations resulted to be not adequate for the re-ranking purpose. In fact, the considered lower-bounds approximate well the actual distance $d(o,q)$ only if o and q are very close to each other in the original metric space, or if $\Gamma_{o,q}$ contains at least one pivot that is close to q and far to o (or vice versa). However, for randomly selected pivots in high dimensional space this is unlikely to happen: for a random pivots p and for an object o not so close to q, we often have that the distances $d(o,p)$ and $d(q,p)$ are both close to the mean value in the distribution of the data distances, and so the lower-bound results to be close to zero. This means that, when using the Pivoted lower-bound for the re-ranking, many objects may be incorrectly swapped and far objects can be assigned in top-positions. More generally, we observed that the Pivoted distance bounds have high relative errors with respect to the actual distance and that these errors slightly decrease when increasing the number h of pivots used to compute the bounds (Fig. 3a).

The Simplex distance bounds showed similar drawbacks when using relatively small prefix lengths. In particular, they are mostly influenced by the fact that the Simplex bounds asymptotically approach the true distances when increasing the number h of pivots used to build the simplex base and that the tightness of the bounds highly depends on h. In fact, in all the cases we observed that there exists a value of $\tilde{h}$ for which full convergence is achieved: 4096 for YFCC100M, around 100 for Twitter-GloVe and SISAP colors. Therefore, for two objects $o,s \in CandSet(q)$ with $|\Gamma_{o,q}| < |\Gamma_{s,q}| \ll \tilde{h}$ we may have

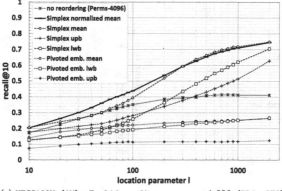

(a) YFCC100M (1M), Euclidean distance, $n = 4,096$ (IDim=278)

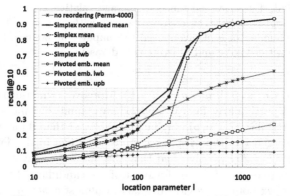

(b) Twitter-GloVe, Cosine distance, $n = 4,000$ (IDim=105)

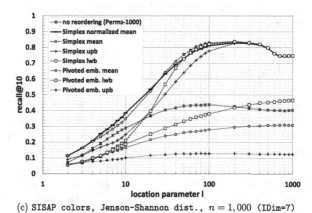

(c) SISAP colors, Jenson-Shannon dist., $n = 1,000$ (IDim=7)

Fig. 2. *Recall*@10 varying the location parameter l (the number n of pivots is fixed). The candidate set to be reordered is selected with a 100-NN search in the permutation space using the Spearman's rho with location parameter l. In the captions we also report the Intrinsic Dimensionality (IDim), computed as in [8], for each considered metric space.

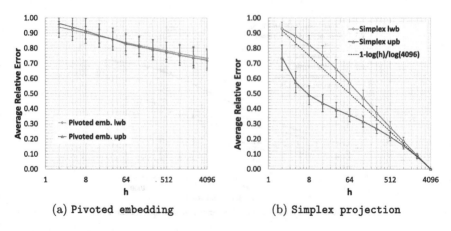

Fig. 3. YFCC100M, Euclidean Distance - Average relative error of the `Pivoted embedding` and `Simplex embedding` distance bounds with respect to the actual distance varying the number of h of pivots used to compute the bounds. Similar trends are obtained on Twitter-GloVe and SISAP colors datasets.

$S_{lwb}(o, q) < S_{lwb}(s, q)$ even if $d(o, q) > d(s, q)$. Moreover, when using few pivots the upper-bound particularly fails in approximate small distances (it is not a metric and in particular $S_{upb}(o, o)$ may be much greater than 0 for small h). The effect of the convergence of the `Simplex` bounds is evident in the Twitter-GloVe data (Fig. 2b): for $l \geq 200$ we observed that the number of pivots in the intersection $\Gamma_{o,q}$ starts to exceed $\tilde{h} = 100$ for most of the candidate objects o, thus in that case all `Simplex` bounds provide an exact or almost exact approximation of the actual distances. A similar phenomenon was observed on the SISAP colors.

In general, we experimentally observed that very good re-ranking scores can be obtained using the same simplex base (e.g. the one formed by the pivots in the query permutation prefix) to project all the candidate objects. However, this is not feasible because at query time, for each candidate object o, we had access only to the distances $d(o, p)$ with p appearing in both the object and query permutation prefixes. Thus, the simplex base used (and its dimension h) changes when considering different candidate objects. This means that the "quality" (tightness) of the `Simplex`-based approximations of the distances query-objects is not uniform within the set of the candidate objects. To overcome this issue, we tested normalized versions of the `Simplex` distance bounds that take into account the number of pivots used for projecting the data. In Fig. 2, we report the normalized mean that was the one obtaining the best results. As normalization factor we used $g(h) = \log(h)$ since we experimentally observed that the relative errors decrease logarithmically with h (e.g. Fig. 3b).

In all the tested cases, the re-ranking using the `Simplex normalized mean` improved the permutation-based results. For example, using about 4, 000 pivots and $l = 300$ the $recall@10$ is improved from 0.39 to 0.59 on YFCC100M dataset, and from 0.43 to 0.76 on Twitter-GloVe. On SISAP colors the recall increase from

0.44 to 0.80 for $l = 80$ and $n = 1,000$. We provided examples with $l < n$ since when using inverted files the number of index blocks accessed is proportional to l^2/n; moreover, it does not depend on the number of retrieved objects.

The disk space needed by the inverted file can be estimated in general assuming to encode each entry of the posting lists with $\lceil log_2|X|\rceil + 32$ bits, where $|X|$ is the size of the dataset. This space is largely sufficient to encode both the ID of the object and its distance from the pivot corresponding to the list to which the entry belongs to. As observed in [4], the positions of the objects can be neglected by ordering the entries of the posting list according to the position of the objects. So, for a fixed l, the size of our index is $l|X|(\lceil log_2|X|\rceil + 32)$ bits, e.g. 1.8 GB for indexing one million objects using $l = 300$. For reference, we observe that for the same set-up the size of the traditional permutation-based inverted index is 0.70 GB. However, more efficient approaches to compressing our posting lists might be employed, for example by quantizing the distances to store each of them in less than 32 bits. In facts, preliminary tests on a 1M subset of YFCC100M deep features, confirmed us that the retrieval performance is preserved when using just 8 bits for storing each distance. In such case, the size of our index is 0.98 GB for one million objects and $l = 300$. For lack of space, we reserve further investigation of this aspect for future work.

Finally, we observe that for $l = 300$ the time cost at query time for computing all the simplex bases and projecting both the query and the candidate objects is about 300 ms. As future work, we plan to reduce the query runtime cost by optimizing the construction of the simplex bases, e.g. by re-using some of them instead of computing a new simplex base from scratch for each candidate object.

5.3   Conclusions

In this article, we presented an approach that proposes to exploit the n-Simplex projection to reorder the candidate list of an approximate k-NN-based search system based on permutations, without accessing the original data. However, our approach can be generalized to other types of approximate search provided that they are based on the use of anchor objects from which we must pre-calculate the distances for other purposes. For example, some data structures use inverted indices, as the inverted multi-index [6], in which objects belonging to a Voronoi cell are inserted in a posting list associated with the centroid of the cell from which we calculated the distance. Other indexes that can benefit from our approach are those based on permutation prefix trees, like PP-Index [13] and PPP-Index [17]. We also intend to investigate developments of our approach using the aggregation of the rankings provided by the permutation representations and the rankings obtained with various n-Simplex bounds, using techniques as that proposed in [17], instead of just re-ranking the former one.

Acknowledgements. The work was partially funded by Smart News, "Social sensing for breaking news", CUP CIPE D58C15000270008, and by VISECH, ARCO-CNR, CUP B56J17001330004.

References

1. Amato, G., Falchi, F., Gennaro, C., Rabitti, F.: YFCC100M-HNfc6: a large-scale deep features benchmark for similarity search. In: Amsaleg, L., Houle, M.E., Schubert, E. (eds.) SISAP 2016. LNCS, vol. 9939, pp. 196–209. Springer, Cham (2016). https://doi.org/10.1007/978-3-319-46759-7_15

2. Amato, G., Falchi, F., Gennaro, C., Vadicamo, L.: Deep permutations: deep convolutional neural networks and permutation-based indexing. In: Amsaleg, L., Houle, M.E., Schubert, E. (eds.) SISAP 2016. LNCS, vol. 9939, pp. 93–106. Springer, Cham (2016). https://doi.org/10.1007/978-3-319-46759-7_7

3. Amato, G., Falchi, F., Rabitti, F., Vadicamo, L.: Some theoretical and experimental observations on permutation spaces and similarity search. In: Traina, A.J.M., Traina, C., Cordeiro, R.L.F. (eds.) SISAP 2014. LNCS, vol. 8821, pp. 37–49. Springer, Cham (2014). https://doi.org/10.1007/978-3-319-11988-5_4

4. Amato, G., Gennaro, C., Savino, P.: MI-File: using inverted files for scalable approximate similarity search. Multimed. Tools Appl. **71**(3), 1333–1362 (2014)

5. Amato, G., Savino, P.: Approximate similarity search in metric spaces using inverted files. In: Proceedings of InfoScale 2008, pp. 28:1–28:10. ICST (2008)

6. Babenko, A., Lempitsky, V.: The inverted multi-index. In: Proceedings of CVPR 2012, pp. 3069–3076. IEEE (2012)

7. Blumenthal, L.M.: Theory and Applications of Distance Geometry. Clarendon Press, Oxford (1953)

8. Chávez, E., Figueroa, K., Navarro, G.: Effective proximity retrieval by ordering permutations. IEEE Trans. Pattern Anal. Mach. Intell. **30**(9), 1647–1658 (2008)

9. Connor, R., Cardillo, F.A., Vadicamo, L., Rabitti, F.: Hilbert exclusion: improved metric search through finite isometric embeddings. ACM Trans. Inf. Syst. **35**(3), 17:1–17:27 (2016)

10. Connor, R., Vadicamo, L., Cardillo, F.A., Rabitti, F.: Supermetric search with the four-point property. In: Amsaleg, L., Houle, M.E., Schubert, E. (eds.) SISAP 2016. LNCS, vol. 9939, pp. 51–64. Springer, Cham (2016). https://doi.org/10.1007/978-3-319-46759-7_4

11. Connor, R., Vadicamo, L., Cardillo, F.A., Rabitti, F.: Supermetric search. Inf. Syst. (2018). https://doi.org/10.1016/j.is.2018.01.002. https://www.sciencedirect.com/science/article/pii/S0306437917301588

12. Connor, R., Vadicamo, L., Rabitti, F.: High-dimensional simplexes for supermetric search. In: Beecks, C., Borutta, F., Kröger, P., Seidl, T. (eds.) SISAP 2017. LNCS, vol. 10609, pp. 96–106. Springer, Cham (2007). https://doi.org/10.1007/978-3-319-68474-1_7

13. Esuli, A.: Use of permutation prefixes for efficient and scalable approximate similarity search. Inf. Process. Manag. **48**(5), 889–902 (2012)

14. Fagin, R., Kumar, R., Sivakumar, D.: Comparing top k lists. In: Proceedings of SODA 2003, pp. 28–36. Society for Industrial and Applied Mathematics (2003)

15. Figueroa, K., Navarro, G., Chávez, E.: Metric spaces library (2007). www.sisap.org/library/manual.pdf

16. Micó, M.L., Oncina, J., Vidal, E.: A new version of the nearest-neighbour approximating and eliminating search algorithm (AESA) with linear preprocessing time and memory requirements. Pattern Recogn. Lett. **15**(1), 9–17 (1994)

17. Novak, D., Zezula, P.: PPP-codes for large-scale similarity searching. In: Hameurlain, A., Küng, J., Wagner, R., Decker, H., Lhotska, L., Link, S. (eds.) Transactions on Large-Scale Data- and Knowledge-Centered Systems XXIV. LNCS, vol. 9510, pp. 61–87. Springer, Heidelberg (2016). https://doi.org/10.1007/978-3-662-49214-7_2

18. Pennington, J., Socher, R., Manning, C.D.: GloVe: global vectors for word representation. Proc. EMNLP **2014**, 1532–1543 (2014)

19. Pestov, V.: Indexability, concentration, and VC theory. J. Discret. Algorithms **13**, 2–18 (2012)

20. Schoenberg, I.J.: Metric spaces and completely monotone functions. Ann. Math. **39**(4), 811–841 (1938)

21. Thomee, B., et al.: YFCC100M: the new data in multimedia research. Commun. ACM **59**(2), 64–73 (2016)

22. Weber, R., Schek, H.-J., Blott, S.: A quantitative analysis and performance study for similarity-search methods in high-dimensional spaces. VLDB **98**, 194–205 (1998)

23. Zezula, P., Amato, G., Dohnal, V., Batko, M.: Similarity Search: The Metric Space Approach. Advances in Database Systems, vol. 32. Springer, Boston (2006). https://doi.org/10.1007/0-387-29151-2

24. Zhou, B., Lapedriza, A., Xiao, J., Torralba, A., Oliva, A.: Learning deep features for scene recognition using places database. In: Advances in Neural Information Processing Systems 27, pp. 487–495. Curran Associates Inc. (2014)

Performance Analysis of Graph-Based Methods for Exact and Approximate Similarity Search in Metric Spaces

Larissa Capobianco Shimomura[1]([✉]), Marcos R. Vieira[2], and Daniel S. Kaster[1]

[1] University of Londrina, Londrina, PR, Brazil
laricsh@gmail.com, dskaster@uel.br
[2] Hitachi America Ltd, R&D, Santa Clara, CA, USA
marcos.vieira@hal.hitachi.com

Abstract. Similarity searches are widely used to retrieve complex data, such as images, videos, and georeferenced data. Recently, graph-based methods have emerged as a very efficient alternative for similarity retrieval. However, to the best of our knowledge, there are no previous works with experimental analysis on a comprehensive number of graph-based methods using the same search algorithm and execution environment. In this work, we survey the main graph-based types currently employed for similarity searches and present an experimental evaluation of the most representative graphs on a common platform. We evaluate the relative performance behavior of the tested graph-based methods with respect to the main construction and query parameters for a variety of real-world datasets. Our experimental results provide a quantitative view of the exact search compared to accurate setups for approximate search. These results reinforce the tradeoff between graph construction cost and search performance according to the construction and search parameters. With respect to the approximate methods, the Navigable Small World graph (*NSW*) presented the highest recall rates. Nevertheless, given a recall rate, there is no winner graph for query performance.

1 Introduction

A wide range of modern applications require similarity retrieval of complex data, such as images, videos, and georeferenced data. A complex data is commonly represented through a feature vector describing to some extent its intrinsic characteristics. The similarity between two complex data is usually measured by applying a distance function to their respective feature vectors [4]. The combination between feature space and distance function is called similarity space, which can be modeled as a metric space [6].

However, similarity searches can have a high computational cost, due to the large volume of data, and a high complexity of calculating similarity functions. In order to speed up similarity queries a large number of access methods has been

This work has been supported by the Brazilian agencies CAPES and CNPq.

S. Marchand-Maillet et al. (Eds.): SISAP 2018, LNCS 11223, pp. 18–32, 2018.
https://doi.org/10.1007/978-3-030-02224-2_2

proposed in the literature. Recently, graph-based methods have emerged as a very efficient option to evaluate similarity queries, however, there is no related work that presents a comprehensive review of graph-based methods.

Survey articles by Chávez et al. [6] and Hjaltason and Samet [12] describe search in metric spaces and cover indexing methods used for similarity searches. The work of Skopal and Bustos [25] provides a broad view of works addressing searches using non-metric similarity functions for different complex data domains. Nevertheless, none of these works include graph-based methods. Graphs are only referenced to introduce theoretical foundations or to present applications on graph datasets. In our work we do not cover such applications since our focus is to use graphs to represent the similarity space itself.

In order to perform similarity searches, graph-based methods require defining what the vertices and edges represent in the graph [18,21,23]. A common approach is to use vertices to represent complex data, and edges to connect two vertices as the similarity relationship between the pair of complex data objects. Notice that in this approach the type of graph is defined by how the vertices are connected. A recent survey by Naidan et al. [19] covered permutation-based methods for similarity search and presented an experimental analysis of some methods used for similarity searches. The methods tested are the graphs Small World Graph (SWG) [18] and the Approximate k-NN Graph built with the NN-Descent [8]. However, the above survey neither explores the impact of different parameters nor addresses exact search in graph-based methods.

In our work we survey the main graph-based methods for exact and approximate similarity search in metric spaces, as well as present an extensive evaluation of representative approaches. We present a comparison between exact and approximate search algorithms in different graph types and analyze how the main parameters affect the performance for both the construction and search algorithms. To avoid too specific findings, we employ two search algorithms: one for exact search and one for approximate search. We then apply them in the most common used graphs for similarity search available in the literature: the k-NN Graph with two construction algorithms, brute-force and NN-Descent; the Relative Neighborhood Graph (RNG); and the Navigable Small World graph (NSW). We also test the Spatial Approximation Tree (SAT) [20] as a baseline for exact search.

We use a wide range of real datasets that vary in dimensionality and cardinality to provide a detailed experimental analysis in a uniform platform. To the best of our knowledge, our work is the first to cover an analysis of exact and approximate search on different graph types for similarity search in metric spaces. The results presented in this paper provide a quantitative view of exact and approximate search, as well as the tradeoff between graph construction cost and search performance according to different construction and search parameters.

This paper is organized as follows: Sect. 2 provides a background of graph-based methods; Sect. 3 covers exact and approximate graph searches; Sect. 4

details the main similarity search graph-based methods; Sect. 5 presents our experimental results; and Sect. 6 concludes this paper.

2   Proximity Graphs and Spatial Approximation

A graph is defined as $G = (V, E)$, where V is the set of vertices (nodes) and E is the set of edges that connect pairs of vertices V. The most common type of graphs used for complex data retrieval using similarity search are the **Proximity Graphs** [18,21,23]. A proximity graph is a graph in which each pair of vertices $(u, v) \in V \times V$ is connected by an edge $e = (u, v), e \in E$, if and only if u and v fulfill a defined property P [21]. Property P is called *neighborhood criterion* and it defines the type of the proximity graph. The edges in E can be weighted or not. The weight of the edge is usually the proximity measurement between the connecting vertices, such as the distance between them $(\delta(u, v))$ [23].

The fundamental approach to perform similarity queries on a proximity graph is to employ the *spatial approximation*, introduced by Navarro in [20]. Given a query element q and proximity measure δ, this approach consists of starting from a given vertex $u \in V$ and iteratively traverse the graph through the "neighbors" of u ($N(u)$: adjacent vertices of u) in a way to get spatially closer to the elements that are the most similar to q. The trivial graph that the spatial approximation approach guarantees exact results for metric spaces is the complete graph (i.e., every vertex is connected to all vertices in the graph). However, in this case, the search would end up in a sequential scan. Therefore, it is necessary to use graphs with fewer edges.

Given a metric space $\langle \mathbb{S}, \delta \rangle$ and a graph $G = (V, E)$, where $V \subseteq \mathbb{S}$ and every $e \in E$ has the form $e = (u, v)$ such that $v \in N(u)$, G must fulfill the Property 1 [20] to correctly answer similarity queries using a search algorithm based on the spatial approximation approach, for any query element $q \in \mathbb{S}$:

$$\forall u \in V, \text{ if } \quad \forall v \in N(u), \delta(q, u) \leq \delta(q, v), \text{ then } \quad \forall v' \in V, \delta(q, u) \leq \delta(q, v') \quad (1)$$

Property 1 means that if it is not possible to get closer q than u through its neighbors, then u is the closest element to q in the graph. Navarro et al. [20] showed that the definition of the Voronoi regions is correspondent to the spatial approximation property. Hence, using the Delaunay Graph (referred to as DG in this paper) – a dual to the Voronoi Diagram [3] – as a structure for the similarity search, search algorithms based on the spatial approximation can return exact results. However, DG has two major limitations: (1) since extra information on the internal structure of the metric space it requires, it is not possible to compute the DG for a generic metric spaces given only the set of dataset distances [18,20,21]; and (2) despite previous works that extend the DG for n-dimensional data [3], the existing solutions suffer from the curse of dimensionality [18], as the graph can become a complete graph very quickly as the dimensionality increases [11]. Therefore, to be more compelling in answering similarity queries using graphs, previous works focused on how to use graphs with a minimized number of edges.

Besides the DG, other proximity graphs are presented in the literature as subgraphs of DG. The hierarchical relation of the most representative graph types are: NNG(Nearest Neighbor Graph) $\subseteq$ RNG(Relative Neighborhood Graph) $\subseteq$ GG(Gabriel Graph) $\subseteq DG$ [21]. These graphs have common edges with the DG for the same vertex set. However, the lacking edges and the edges that do not exist in the DG can lead a search algorithm based on spatial approximation to a local approximate answer.

To solve similarity searches Navarro et al. [20] proposed the Spatial Approximation Tree (SAT). SAT is a particular case of a DG subgraph in which the resulting structure is a tree with a fixed root to start the search, which returns exact results. Ocsa et al. [21] stated that the selection of a bad root can lead to excessive node exploration. They also showed that an RNG can reduce significantly the number of distance computations in queries when compared to SAT.

3   Exact and Approximate Search in Proximity Graphs

Search algorithms for graph-based methods with exact and approximate results were proposed. For exact search in metric spaces, Paredes and Chávez [23] proposed algorithms for k-NN and $Range$ queries using weighted directed k-NN graphs. Their main contribution is to use weighted k-NNG to estimate lower and upper bounds between query and dataset elements. These bounds and other pruning strategies are used to discard elements (vertices) in the dataset that are not in the result set according to **metric space properties**. As a result, this algorithm may not give exact result answers in general non-metric spaces.

Because of the complexity of similarity searches, a large amount of efforts has concentrated in similarity search algorithms with approximate results. For example, a strategy used to search over k-NN graphs is to select initial vertices and use a greedy best-first search process based on the spatial approximation [2].

The Greedy Search algorithm (GS) [1] starts by randomly choosing a vertex of the graph as the start vertex v_i then, it computes the similarity between the query q and each neighbor of v_i; the neighbor vertex with the highest similarity to q is then selected to be the next v_i. This process continues until there are no neighbors with higher similarity value to q than v_i.

The $GNNS$ [10] can be seen as an extension of the GS algorithm. This algorithm starts with a randomly chosen initial vertex $u \in V$ and, in each iteration, u is swapped with $v \in V$, where v is the neighbor of u that is most similar to q. This process ends after a maximum number of swaps T. The above greedy search process is restarted R times with different starting vertex. The final result is the set of k most similar vertices to q evaluated in all random restarts R.

The major drawback of search algorithms based on spatial approximation is when the starting node is "far away" from the result, thus leading to a large running time that depends on the graph size. To minimize the execution time as well as to increase the probability of finding the correct answer, *seed nodes* are selected as starting vertices. These seed nodes can be selected by random sampling, e.g., using a second index structure [26] or selecting well-separated elements in a dataset [2].

4   Graph-Based Methods for Similarity Searching

Graphs that fulfill the spatial approximation property (e.g., complete graph and DG) have critical limitations, as previously discussed. Hence, a large amount of research has focused on using graphs with a minimized number of edges for **approximate similarity search** [18]. In this section, we describe the main types of graphs currently used for similarity search.

***k-NN* Graphs.** The *k-NNG* [23] is defined as a graph $G = (V, E)$, where $E = \{(u, v, \delta(u, v)) | v \in NN_k(u)_\delta\}$ such that $NN_k(u)_\delta$ is the set containing the k nearest neighbors of u in the set of vertices V using the similarity function δ. The edges of *k-NNG* can be undirected or directed (*k-NNG_u*).

The trivial brute force algorithm for constructing *k-NNG* (denoted *k-NNG-BF*) has cost $O(n^2)$. Thus, some research works focused on developing better algorithms for *k-NNG* construction [8]. Paredes et al. [24] proposed two algorithms for *k-NNG* construction, the Recursive Partition Based Algorithm and the Pivot Based Algorithm. Although these two methods presented good results in the experiments when compared to the *k-NNG-BF* algorithm, both methods require the construction of secondary global structures to make the $NN_k(u)_\delta$ computation faster when building the *k-NNG*. A second proposed *k-NNG* construction algorithm is the *NN-Descent* algorithm [8]. In this approach, an approximate *k-NNG* is built with arbitrary similarity measures, without the need of a secondary global structure, and with a lower computational cost. The basic idea of *NN-Descent* algorithm is "the neighbor of a neighbor" is also probably a "neighbor". It starts by first computing a random approximation of *k-NN* for each element in V, and then iteratively improve the first approximation comparing each element to its neighbor's neighbors. Besides the above methods, other approaches build either exact or approximated *k-NNG* with lower complexity than *k-NNG-BF* for specific metric spaces or general similarity functions [27].

There are other proposed graph-based methods in the literature that are either built from a *k-NNG* or has the same properties as *k-NNG*. One variation of the *k-NNG* is the *k-Degree Reduced Graph* (*k-DRG*) [1]. As the name suggests, the main idea of the *k-DRG* is to build a graph from the *k-NNG* with a *reduced number* of edges that guarantees the k nearest neighbors of a vertex can be reached using the *GS* algorithm. Additional *k-NNG* variations include the *quasi-proximity graphs* [7] and the Pruned Bi-Directed k-Nearest Neighbor Graph (*PBKNNG*) [13].

Relative Neighborhood Graph. The Relative Neighborhood Graph (*RNG*) is a subgraph of DG with guarantees that for any pair of vertices there exist one or more paths that connects a pair of vertices (i.e., *RNG* is a connected graph [21]). Formally, the *RNG* is a graph $G = (V, E)$ whose set of edges E is determined by a proximity property P that $(u, v) \in E$ if and only if $B_u(u, \delta(u, v)) \cap B_v(v, \delta(v, u)) = \emptyset$, where $u, v \in V$ and $B(x, r)$ is a ball defined by x and distance threshold (radius) r [21]. A very important property of *RNG* is that it is parameter-free, i.e., the amount of edges in the graph is "auto-adjusted" as it is defined by the dataset properties.

Distinctly from DG, RNG can be built given only the distances between the dataset elements. However, the complexity to build RNG in general spaces is $O(n^3)$, which is even higher than the complexity of k-NNG-BF. Methods to construct RNG with lower complexity were proposed, however, they require to have a DG in order to then build RNG [14]. In our experiments we used the brute force algorithm ($O(n^3)$ complexity) to build the RNG, as there are limitations to build DG for high dimensional data. Variations of RNG include the hierarchical structure, Hyperspherical Region Graph (HRG) and its improved version MOBHRG [9].

Navigable Small World Graph. Recently, new types of graphs have been proposed for approximate similarity search. Representatives of this type of graph are the Navigable Small World graph (NSW) [17,18], the Randomized Neighborhood graph [2] and the Fast Approximate Nearest Neighbor graph [11]. These proposed graphs achieved high recall rates and performed reasonably well in high dimensional metric spaces, as reported in [11,18].

A Small World Graph (SWG) is a random graph in which pair of vertices is connected by a path that is considered *small* compared to the size of the graph. The proposed NSW graph is based on an approximation of DG and the SWG topology. Vertices of the NSW are connected by two types of edges: *short-range links*, used for the greedy search, and *long-range links*, which define the *small world properties* in the graph.

The NSW construction algorithm is based on iteratively inserting new vertices to the graph. Given a new vertex, a search in the current graph is performed to identify the location and to connect its k nearest neighbors to the newly inserted vertex. The newly created edge is defined as *short-range link* since the edges connect the new vertex to its current k nearest neighbors. In each iteration the edge that was a *short-range link* becomes a *long-range link*.

5   Experimental Evaluation

This section presents an extensive experimental evaluation of graph-based methods for similarity search in metric spaces. We employed the C++ library NMSLib (Non-Metric Space Library) [5] to develop all tested methods. The library originally includes NN-$Descent$ and NSW. Therefore, we extended it by implementing Brute Force k-NNG and RNG. We also implemented the approximate search algorithms GS, $GNNS$ to ensure that every structure employs exactly the same code for searching. We implemented the exact search algorithm using weighted versions of RNG and k-NNG, denoted herein as $RNGW$ and k-$NNGW$. We also tested the SAT implementation available in NMSLib. Table 1 shows a summary of the evaluated methods.

Table 2 shows the six datasets we employed in our analysis and their respective size and dimensionality. From each dataset, we removed 100 random elements to serve as query elements and used the remaining ones to build the

Table 1. List of evaluated methods.

Method	Search algorithm	Construction algorithm
k-NNG	Approximate (*GS* and *GNNS*)	Brute-force $O(n^2)$ (denoted as *k-NNG-BF*) and *NN-Descent* (approximate *k-NNG*)
RNG	Approximate (*GS* and *GNNS*)	Brute-force $O(n^3)$
NSW	Approximate (*GS* and *GNNS*)	Consecutive vertex insertion with edges based on approximate *k-NN* search
RNGW	Exact [23]	Brute-force $O(n^3)$
k-NNGW	Exact [23]	Brute-force $O(n^2)$
SAT	Exact [20]	Consecutive node insertion $O(\frac{n \log^2 n}{\log \log n})$

graphs. For the ANN-SIFT1M we used only the base vectors, the query set was not employed. We employed the Euclidean distance (L_2) as the similarity measure for all methods. Such wide combination of search methods and graphs implemented in a common platform allows us to study the performance behavior of several factors of the graph structure, search method and parameters, in order to understand the inherent tradeoffs among methods. We used the average query time, number of distance computations and recall (fraction of correct query answers retrieved) to evaluate search performance. The experiments were carried out on an Intel Core i5 (4 GB RAM) with a single thread for all methods on an Ubuntu GNU/Linux 16.04 64 bits.

Table 2. Datasets used for the experiments.

Dataset	Size	Dim.	Source
USCities	25,374	2	Geographic coordinates from American cities[a]
Color Moments	68,040	9	Feature vector from *Corel* [22]
Texture	68,040	16	Co-ocurrence texture descriptor from *Corel* [22]
Color Histogram	68,040	32	Color histogram descriptor from *Corel* [22]
MNIST	70,000	784	Images of handwritten digits [16][b]
ANN-SIFT1M (base vectors)	1,000,000	128	SIFT descriptors [15]

[a]http://www.census.gov/main/www/cen2000.html
[b]http://yann.lecun.com/exdb/mnist

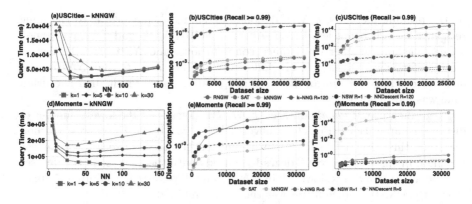

Fig. 1. Performance comparison between exact and approximate searches with recall ≥ 0.99 and 10-NN for *USCities* and *Moments* with different values for *NN* (a–d) and dataset size (b–d and e–f).

5.1 Exact Search Evaluation

This section discusses how proximity graphs behave to produce exact answers for *k-NN* similarity queries. We first analyze how the performance of the exact algorithm in *k-NNGW* is affected by the number of neighbors used to build the graph, denoted as *NN*. The first column of Fig. 1 shows the average query time to answer *k-NN* queries, for $k \in \{1, 5, 10, 30\}$, in *USCities* and *Moments* datasets, for increasing *NN* values. It can be noticed that in general the query time drops abruptly as the *NN* parameter increases, and after reaching the best *NN* value it grows again. Moreover, different values for *k* in queries demand different *NN* to achieve the best performance. Regarding this aspect the dataset properties are of major impact. The best query times for different *k* values in *USCities* (Fig. 1(a)) are achieved for directly proportional values for *k* and *NN* (i.e., bigger performance is achieved for larger *k* values). In contrast, for *Moments* (Fig. 1(d)), such a relation is inversely proportional as, for example, for $k = 30$ the best *NN* is 25, and for $k = 1$ the best *NN* is 150.

We also compare the performance of the exact algorithms using *k-NNGW*, *RNGW* and *SAT* with the performance of the approximate *GNNS* algorithm with configurations of *k-NNG-BF*, *NN-Descent* and *NSW* that provide highly accurate results (recall of at least 0.99) in the smallest query time for 10-NN queries. The second and third columns of Fig. 1 show, respectively, the average number of distance computations and query time regarding the datasets, both in $\log_{10}$ scale. All graphs employed $NN = 100$ (except *RNGW* that is parameter free). The best number of restarts for the *GNNS* are denoted as *R* in the figure.

For the *USCities* dataset, it is noticeable that the *RNG* exact algorithm performed significantly fewer distance computations than the other methods (Fig. 1(b)). However, regarding query time (Fig. 1(c)), the behavior is the opposite, being that *RNGW* is the slowest option by orders of magnitude. This huge performance degradation is due to internal computations of the algorithm used by the pruning methods. *k-NNGW* follows a similar behavior due to the same reason, although being less extreme than *RNGW*. *NSW* and *SAT* performed an intermediate number of distance computations, and *NSW* was the fastest method for this dataset, closely followed by *SAT*. *k-NNG* and *NN-Descent* demanded a huge amount of distance computations, as a result their execution time was intermediate.

Regarding the *Moments* dataset, *k-NNGW* demanded the lowest number of distance computations (Fig. 1(e)), nevertheless it was by far the slowest method regarding time (Fig. 1(f)). *RNGW* was omitted because its execution time was too high. In this dataset, *SAT* degraded quickly with the increase in the dataset size, being up to 3 times slower than *k-NNG* and *NN-Descent*, and up to 5 times slower than *NSW*, which again was the fastest method.

It is worth mentioning that this test employed the cheap Euclidean metric. These results confirm that for cheap metrics, approximate search is much more feasible for graphs, when it is enough to provide answers close to the exact ones. However, the exact algorithm may be competitive for costly metrics.

5.2   Analysis of Construction Scalability and Parameters

Aside from *RNG* and *SAT*, all other tested methods have parameters to tune. Table 3 shows the construction parameters evaluated in this section, the tested values and their description. To limit the number of variations, η and w were set to defaults. After having constructed the graphs, we executed batteries of 1-NN queries using the *GS* algorithm.

Table 3. Construction parameters tested.

Graph	Parameter	Tested values	Description
k-NNG-BF, *NN-Descent* and *NSW*	*NN*	{5, 10, 25, 40, 55, 70, 100, 130, 150}	Number of neighbors per vertex (*k* for *k-NN* searches in *NSW*)
NN-Descent	ρ	{1.0, 0.5}	Sample rate of neighbor's neighbors to be checked for each vertex
NSW	*efConstruction*	{20, 100}	Size of the candidate list (*ef*) of the *k-NN* search in vertex insertion
NN-Descent	η	0.001 (default)	Early termination threshold
NSW	w	1 (default)	Number of starts for *k-NN* search

Figure 2 presents the results (construction time and query time plots are in $\log_{10}$ scale). It is visible in plots (a)–(c) in the figure that both *NSW* and *NN-Descent* demand more indexing time as the *NN* increases for all datasets.

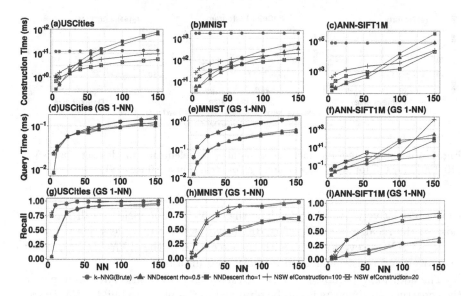

Fig. 2. Construction time (first row), query time (second row), and recall (third row) for increasing *NN* values.

In contrast, our *k-NNG-BF* implementation executes a sequential scan over the dataset to identify the *k*-neighbors of each vertex and, consequently, the variation in the construction time with the increase of *NN* is insignificant. Even though *NN-Descent* method has lower complexity than the brute force algorithm, for large *NN* *NN-Descent* takes close to longer time to build the graph if compared to *k-NNG-BF*. The exception, in our tests, is the *MNIST* high dimensional dataset that even for *NN* = 150 was still faster than *k-NNG-BF*, nevertheless presenting a monotonically increasing pattern. *NSW* demands more time than *NN-Descent* to be constructed for small *NN*, however, the behavior for large *NN* is the opposite. Even though *NN* increases the construction time in *NSW* because of the search process in vertices insertion, a large *NN* has more impact in *NN-Descent* as it increases the number of neighbors to compare in each iteration. Notice that both construction time and query time for *ANN-SIFT1M* were degraded by swap operations for large *NN*, as for some configurations the available RAM was exhausted.

Both the query execution time (Fig. 2(d)–(f)) and recall rate (Fig. 2(g)–(i)) consistently rise with the increase of the *NN* parameter for all datasets. As a role of thumb, the more precise a method is the more query time it demands. If the *NN* parameter is large, the constructed graph is more connected and *GS* has greater chances of returning the correct query answer regardless of the starting vertex. However, the larger the *NN* the more adjacent vertices is evaluated in each greedy step, thus consuming more time. *NSW* was the slowest method in the majority of the cases, however, it was the most precise even for *MNIST* and *ANN-SIFT1M*, the tested datasets with the highest dimensionality and cardi-

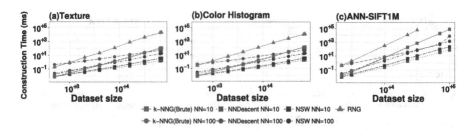

Fig. 3. Indexing time for increasing number of elements in dataset for $NN \in \{10, 100\}, \rho = 0.5$ for $NN\text{-}Descent$ and $efConstruction = 20$ for NSW

nality, respectively. Another result is that $NN\text{-}Descent$ yields recall rates very close to the recall rates of $k\text{-}NNG\text{-}BF$, however demanding less query time for the same NN in several cases, with the exception of the $ANN\text{-}SIFT1M$ due to swap overhead consequent from memory consumption.

The additional parameters of $NN\text{-}Descent$ and NSW also have an impact on the construction time. As it can be observed in the plots, to build a $NN\text{-}Descent$ with $\rho = 1$ demanded in average 30% more time than to build it for the same dataset and NN value with $\rho = 0.5$. Unexpectedly, such a rise in construction time did not reflect corresponding gains in query time or recall as they were close for both tested ρ values. With regard to NSW, the time elapsed to build a graph for the same dataset and NN value for $efConstruction = 100$ was in average 40% larger than for $efConstruction = 20$. Such a cost increase resulted in higher recall rates for some datasets, and overall negligible query overhead, except for $ANN\text{-}SIFT1M$ in which some configurations suffered from swap. Nevertheless, the dominating parameter, according to our tests, in terms of cost-benefit for NSW is NN. For instance, the time elapsed to build NSW with $NN = 25$ and $efConstruction = 100$ is comparable to the time to build NSW with $NN = 70$ and $efConstruction = 20$ for $MNIST$, however the recall of the second configuration is around 30% superior than that of the first.

We evaluated the scalability of the different graphs construction by increasing continuously the dataset size from a small number of elements to the full dataset. Figure 3 shows the results in $\log_{10}$ scale. Due to time limitations, we did not construct RNG for more than 100,000 elements in $ANN\text{-}SIFT1M$ dataset. As expected, considering the algorithm complexity, RNG takes the longest and also has the worst scalability to construct among the tested methods. The construction time difference is noticeable specially for $ANN\text{-}SIFT1M$ dataset in which RNG construction time for 100,000 elements is just approximately 20% slower compared to a 100-NNG for the entire dataset (10 times more elements). For graphs $k\text{-}NNG$, $NN\text{-}Descent$ and NSW, both with $NN = 10$, when the number of elements doubles, the construction time increases in average 280%, 100% and 130%, respectively.

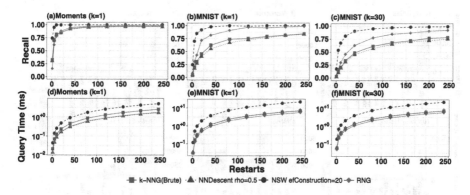

Fig. 4. Recall and query time when increasing number of restarts for moments and MNIST datasets with the construction parameters for *k-NNG*, *NN-Descent* and *NSW* graph: $NN = 10, \rho = 0.5$ for *NN-Descent* and $efConstruction = 20$ for *NSW* graph.

5.3   Analysis of the Number of Restarts in the *GNNS* Search

To improve the result quality of *GS*, we employed the *GNNS* algo-rithm. We varied the number of restarts R of the greedy process in $\{1, 5, 10, 20, 40, 80, 120, 160, 200, 240\}$ and evaluated all the neighbors of a vertex in each greedy step. Each greedy process terminates when the algorithm reaches the best answer (vertex) when compared to its neighbors. Figure 4 shows the recall and query time according to restarts for $k = 1$ and $k = 30$ *k-NN* queries.

As expected, *k-NNG-BF* and *NN-Descent* had similar query time and recall according to the increase of restarts (R). In *Moments* dataset (Fig. 4(a), (d)) *RNG* has similar or better recall than *k-NNG* (Brute force and *NN-Descent*) for 1-NN queries. Overall, *NSW* needs a small number of restarts to achieve high recall rates in lower dimensional datasets. In *Moments* dataset *NSW* needs only 5 restarts to achieve recall $= 1$, while *RNG* required 80 restarts. In this case, the *NSW* also performed better in query time. For 30-NN searches in *MNIST* (Fig. 4(c), (f)) *NSW* and *RNG* achieved approximately the same recall with 40 and 200 restarts, respectively; unexpectedly, with approximate query time.

As we are interested in results as close to exact results as possible, we selected the search parameter settings for each method that achieved recall rates bigger than 0.9 and compared it by distance computations, as this is correlated to the query time in *GNNS* algorithm. The *GNNS* is a best-first algorithm hence the bigger the number of distance computations the longer the query takes to exe-cute. To avoid too many settings we fixed the following construction parameters: $NN \in \{10, 55, 100\}, \rho = 0.5$ and $efConstruction = 20$. The results are in Fig. 5 and the needed number of restarts (R) is displayed in method label. Graphs with the selected parameters that did not achieve recall ≥ 0.9 are not displayed.

To achieve the same recall rate, graphs built with bigger NN parameter val-ues need a small number of restarts. For example in *Texture* (Fig. 5(c)) dataset, *k-NNG*, *NN-Descent* built with $NN = 10$ needs 80 restarts while for $NN = 55$ (around 5 times more connected) only R $= 1$ is enough to achieve 0.9 recall. In

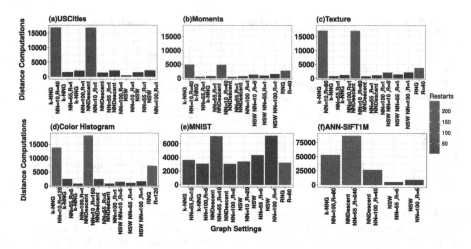

Fig. 5. Distance computations and parameters for *GNNS* search algorithm for *RNG*, *k-NNG*, *NSW* and *NN-Descent* to achieve a recall rate larger than 0.9 for *NN* parameters: [10, 55, 100] for 1-NN search.

cases in which the same number of restarts is needed to achieve a determined recall but the *NN* parameter of the graph is different, the search is faster (fewer distance computations) in the graph with the smaller *NN* parameter. This happens for *k-NN*, *NN-Descent* and *NSW* R $= 1, NN = 55$ and $NN = 100$ in *Moments*, *Texture* and *USCities* datasets.

As already discussed, *NSW* needs a very small number of restarts to achieve high recall rates, as long-range edges can act as shortcuts for data that are far from each other. There are cases that the *NSW* number of computations is comparable to other methods. For example, for $NN = 55$ in *USCities* dataset, *NSW* has approximately the same number of distance computations than *k-NNG* and *NN-Descent*. However, for large datasets such as *ANN-SIFT1M* a *NSW* with the same *NN* outperforms *k-NNG* and *NN-Descent* by far.

Figure 5(b)–(d) shows *RNG* can outperform $NN = 10$ *k-NNG* and *NN-Descent*, but needs more restarts. Even though *RNG* needed 80 restarts in (e), 16 times more than $NN = 100$ *k-NNG*, *NN-Descent* and *NSW* needed, the number of distance computations are comparable. Another interesting point of Fig. 5(e) is that the *NSW* had the worst results compared to *k-NNG* and *NN-Descent* for the same *NN* parameter, with a visible difference for $NN = 100$. The restarts can be tuned to achieve high recall, however, the query time and distance computations needed to achieve it relies on the graph structure.

6   Conclusion and Future Works

In this paper, we surveyed and evaluated the performance of several representative graph-based methods used for exact and approximate similarity search, according to their main construction and search parameters for a range of real-world datasets. From the experimental results we observed that even though the exact search algorithm had fewer distance computations, query execution time is large when compared to approximate search. For k-NNG, NN-$Descent$ and NSW, there is a tradeoff between construction and query time. As the number of neighbors per vertex increases, query time and number of restarts needed to return answers close to exact decreases. Graph NSW outperforms other methods in construction time. However, when comparing graph settings for a given recall rate we were not able to point out a winner method for every condition tested. The results also indicates that RNG query performance can be competitive specially in high dimensional datasets.

As ongoing work, we are exploring costly metrics and analyzing how the datasets distribution affects the graphs performance. The key goal is to provide deep analysis on the tradeoffs between construction time and query time for graph-based structures.

References

1. Aoyama, K., Saito, K., Yamada, T., Ueda, N.: Fast similarity search in small-world networks. In: Fortunato, S., Mangioni, G., Menezes, R., Nicosia, V. (eds.) Complex Networks. Studies in Computational Intelligence, vol. 207, pp. 185–196. Springer, Berlin (2009). https://doi.org/10.1007/978-3-642-01206-8_16
2. Arya, S., Mount, D.M.: Approximate nearest neighbor queries in fixed dimensions. In: SODA, pp. 271–280 (1993)
3. Aurenhammer, F.: Voronoi diagrams - a survey of a fundamental geometric data structure. ACM Comput. Surv. **23**(3), 345–405 (1991)
4. Barioni, M.C.N., dos Santos Kaster, D., Razente, H.L., Traina, A.J., Júnior, C.T.: Advanced Database Query Systems. IGI Global, Hershey (2011)
5. Boytsov, L., Naidan, B.: Engineering efficient and effective non-metric space library. In: Brisaboa, N., Pedreira, O., Zezula, P. (eds.) SISAP 2013. LNCS, vol. 8199, pp. 280–293. Springer, Heidelberg (2013). https://doi.org/10.1007/978-3-642-41062-8_28
6. Chávez, E., Navarro, G., Baeza-Yates, R., Marroquín, J.L.: Searching in metric spaces. ACM Comput. Surv. **33**(3), 273–321 (2001)
7. Chávez, E., Sadit Tellez, E.: Navigating k-nearest neighbor graphs to solve nearest neighbor searches. In: Martínez-Trinidad, J.F., Carrasco-Ochoa, J.A., Kittler, J. (eds.) MCPR 2010. LNCS, vol. 6256, pp. 270–280. Springer, Heidelberg (2010). https://doi.org/10.1007/978-3-642-15992-3_29
8. Dong, W., Moses, C., Li, K.: Efficient K-nearest neighbor graph construction for generic similarity measures. In: Proceedings of WWW, pp. 577–586 (2011)
9. Florez, O.U., Qi, X., Ocsa, A.: MOBHRG: fast k-nearest-neighbor search by overlap reduction of hyperspherical regions. In: ICASSP, pp. 1133–1136 (2009)
10. Hajebi, K., Abbasi-Yadkori, Y., Shahbazi, H., Zhang, H.: Fast approximate nearest-neighbor search with k-nearest neighbor graph. In: IJCAI, pp. 1312–1317 (2011)
11. Harwood, B., Drummond, T.: FANNG: fast approximate nearest neighbour graphs. In: CVPR, pp. 5713–5722 (2016)
12. Hjaltason, G.R., Samet, H.: Index-driven similarity search in metric spaces. ACM Trans. Database Syst. **28**(4), 517–580 (2003)

13. Iwasaki, M.: Pruned bi-directed k-nearest neighbor graph for proximity search. In: Amsaleg, L., Houle, M.E., Schubert, E. (eds.) SISAP 2016. LNCS, vol. 9939, pp. 20–33. Springer, Cham (2016). https://doi.org/10.1007/978-3-319-46759-7_2

14. Jaromczyk, J.W., Toussaint, G.T.: Relative neighborhood graphs and their relatives. Proc. IEEE **80**(9), 1502–1517 (1992)

15. Jegou, H., Douze, M., Schmid, C.: Product quantization for nearest neighbor search. PAMI **33**(1), 117–128 (2011)

16. Lecun, Y., Bottou, L., Bengio, Y., Haffner, P.: Gradient-based learning applied to document recognition. Proc. IEEE **86**(11), 2278–2324 (1998)

17. Malkov, Y., Ponomarenko, A., Logvinov, A., Krylov, V.: Scalable distributed algorithm for approximate nearest neighbor search problem in high dimensional general metric spaces. In: Navarro, G., Pestov, V. (eds.) SISAP 2012. LNCS, vol. 7404, pp. 132–147. Springer, Heidelberg (2012). https://doi.org/10.1007/978-3-642-32153-5_10

18. Malkov, Y., et al.: Approximate nearest neighbor algorithm based on navigable small world graphs. Inf. Syst. **45**, 61–68 (2014)

19. Naidan, B., Boytsov, L., Nyberg, E.: Permutation search methods are efficient, yet faster search is possible. Proc. VLDB Endow. **8**(12), 1618–1629 (2015)

20. Navarro, G.: Searching in metric spaces by spatial approximation. VLDB J. **11**(1), 28–46 (2002)

21. Ocsa, A., Bedregal, C., Cuadros-Vargas, E.: A new approach for similarity queries using neighborhood graphs. In: Brazilian Symposium on Databases, pp. 131–142 (2007)

22. Ortega, M., Rui, Y., Chakrabarti, K., Porkaew, K., Mehrotra, S., Huang, T.S.: Supporting ranked Boolean similarity queries in MARS. TKDE **10**(6), 905–925 (1998)

23. Paredes, R., Chávez, E.: Using the k-nearest neighbor graph for proximity searching in metric spaces. In: Consens, M., Navarro, G. (eds.) SPIRE 2005. LNCS, vol. 3772, pp. 127–138. Springer, Heidelberg (2005). https://doi.org/10.1007/11575832_14

24. Paredes, R., Chávez, E., Figueroa, K., Navarro, G.: Practical construction of k-nearest neighbor graphs in metric spaces. In: Àlvarez, C., Serna, M. (eds.) WEA 2006. LNCS, vol. 4007, pp. 85–97. Springer, Heidelberg (2006). https://doi.org/10.1007/11764298_8

25. Skopal, T., Bustos, B.: On nonmetric similarity search problems in complex domains. ACM Comput. Surv. **43**(4), 1–50 (2011)

26. Wang, J., Li, S.: Query-driven iterated neighborhood graph search for large scale indexing. In: ACM MM, pp. 179–188 (2012)

27. Wang, J., Wang, J., Zeng, G., Tu, Z., Gan, R., Li, S.: Scalable k-NN graph construction for visual descriptors. In: CVPR, pp. 1106–1113 (2012)

Querying Metric Spaces with Bit Operations

Richard Connor[1] and Alan Dearle[2(✉)]

[1] Department of Computing Science, University of Stirling,
Stirling FK9 4LA, Scotland
richard.connor@stir.ac.uk
[2] School of Computer Science, University of St Andrews,
St Andrews KY16 9SS, Scotland
alan.dearle@st-andrews.ac.uk

Abstract. Metric search techniques can be usefully characterised by the time at which distance calculations are performed during a query. Most exact search mechanisms use a "just-in-time" approach where distances are calculated as part of a navigational strategy. An alternative is to use a "one-time" approach, where distances to a fixed set of reference objects are calculated at the start of each query. These distances are typically used to re-cast data and queries into a different space where querying is more efficient, allowing an approximate solution to be obtained.

In this paper we use a "one-time" approach for an exact search mechanism. A fixed set of reference objects is used to define a large set of regions within the original space, and each query is assessed with respect to the definition of these regions. Data is then accessed if, and only if, it is useful for the calculation of the query solution.

As dimensionality increases, the number of defined regions must increase, but the memory required for the exclusion calculation does not. We show that the technique gives excellent performance over the SISAP benchmark data sets, and most interestingly we show how increases in dimensionality may be countered by relatively modest increases in the number of reference objects used.

1 Context

To set a formal context, we are interested in searching a (large) finite set of objects S which is a subset of an infinite set U, where (U, d) is a metric space: that is, an ordered pair (U, d), where U is a domain of objects and d is a total distance function $d : U \times U \to \mathbb{R}$, satisfying postulates of non-negativity, identity, symmetry, and triangle inequality [20]. The general requirement is to efficiently find members of S which are similar to an arbitrary member of U given as a query, where the distance function d gives the only way by which any two objects may be compared. There are many important practical examples captured by this mathematical framework, see for example [16,20]. The simplest type of similarity query is the *range search* query: for some threshold t, based on a query $q \in U$, the solution set is $R = \{s \in S| \, d(q, s) \leq t\}$.

© Springer Nature Switzerland AG 2018
S. Marchand-Maillet et al. (Eds.): SISAP 2018, LNCS 11223, pp. 33–46, 2018.
https://doi.org/10.1007/978-3-030-02224-2_3

The essence of metric search is to spend time pre-processing the finite set S so that solutions to queries can be efficiently calculated using only distances among objects. In all cases therefore, distances between the data and selected reference or "pivot" objects are calculated during pre-processing, and at query time distances between the query and the same pivot objects can be used to make deductions about which data values may, or may not, be candidate solutions to the query.

Mechanisms for metric search can be divided into two main categories, which we define as using "just-in-time" and "one-time" distance calculations between the query and these pivot objects. With "just-in-time" solutions, the manner in which data is stored reflects proximity within the data set, and indexing structures attempt to allow navigation towards subsets where possible solutions may exist. As navigation occurs, distances to objects related to these local subsets are calculated. The idea is that as the computation progresses, subsets of the data which are geometrically distant from the query are never accessed.

With "one-time" solutions, a selection of pre-determined reference objects is used, and distances to all these are calculated for every element during construction. These distances are used to re-cast the original space into some other space where indexing properties are better, distance calculations are cheaper, or both. Typically the main tradeoff is that the extra query efficiency is achieved in return for a loss of semantic query effectiveness, so such mechanisms are either approximate, or produce candidate sets of results from within which the true results must be determined by re-accessing the original data.

2   Introduction

In this paper, we present a combination which we believe is completely novel, and which is robust in the face of increasing dimensionality. We use a "one-time" approach for an exact search mechanism, and thus characterise the original search space by the distances between each element and a fixed set of reference objects. However, instead of using this information to re-cast the data into some cheaper space, we instead assess queries in terms of the original metric space.

Our initial characterisation is used to define a large set of binary partitions over the original space; the data is stored as a set of bitmaps according to containment with these regions. At query time, the query is assessed against all of the regions, but without reference to the data representation. For each region, one of three possible conditions may be determined: (a) the solution set to the query must be fully contained in the region, or (b) there is no intersection between the region and the solution set, or (c) neither of these is the case. In either case (a) or (b), the containment information stored may be useful with respect to solving the query, and will be fetched from disk as part of the solution computation. In case (c) however the containment information is of no value, and is not accessed as a part of the computation. This approach maximises the effectiveness of memory used for calculation against the data representation. Indeed the amount of memory required depends on the size of the data but, critically, not its dimensionality.

As dimensionality increases, then ever more regions will fail to contribute knowledge towards the possible solution set. Since these regions are not involved in the exclusion computation, they do not significantly impact upon the memory or time required to perform it. Thus whilst an increase of dimensionality will require a larger representation of the data in secondary storage to remain effective, it does not increase the memory requirement for performing the exclusion calculation.

2.1 Illustrated Example

Figure 1 shows a simple example within the 2D plane, comprising four reference objects p_1 to p_4 and a set of six regions defined by them. The regions are respectively: A, the area to the left of the hyperplane between p_1 and p_2; B, the area above the hyperplane between p_1 and p_3; and the areas within the variously sized circles drawn around each p_1 to p_4, labelled C, D, E and F respectively. Note that each regional boundary defines a binary partition of the total space, such that each element of the space is either in, or out, of the region, and that this membership is defined only in terms of distances from defined reference objects. Thus in Fig. 1, $A = \{u \in U | d(u, p_1) \geq d(u, p_2)\}, C = \{u \in U | d(u, p_1) \leq \mu\}$ for some value of μ, etc.

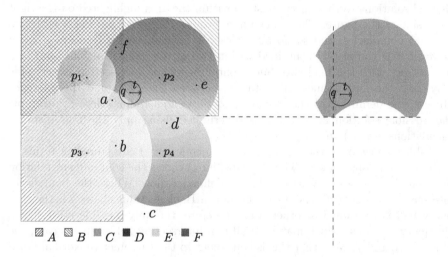

Fig. 1. Any solution to the query q with threshold t must be in the circle centred around p_2, and above the dashed horizontal line; it cannot be in the circles centred around p_1 or p_4. No information is available wrt the circle around p_3 nor the vertical line; these partitions take no part in the exclusion calculation as the region boundaries intersect the query solution boundary. The shaded area shown on the right shows the possible loci of query solutions with respect to these regions. (Color figure online)

Table 1. The data representation according to containment of regions (see Fig. 1). The representation of the query Q is equivalent to the CNF expression $B \wedge \neg C \wedge D \wedge \neg F$

Point	Regions					
	A	B	C	D	E	F
a	True	True	False	True	True	False
b	True	False	False	False	True	True
c	False	False	False	False	False	False
d	False	False	False	True	False	True
e	False	True	False	True	False	False
f	True	True	False	True	False	False
Q	$\cap$	True	False	True	$\cap$	False

Note that the number of regions that can be defined from a fixed set of reference objects is potentially very large; for example in the experiments described in this paper, for n reference objects, we define $\binom{n}{2} + 5n$ regions for n reference objects, by using all hyperplane boundaries and defining five hypersphere radii per object.

Figure 1 also shows a range query q drawn with a threshold t. It can be seen that all solutions to this query must lie within the area highlighted on the right hand side of the figure. The hypersphere around the query intersects with two regional boundaries, and so no information is available with respect to these; however it is completely contained within two of the defined regions, and fails to intersect with the final two. Such containment and intersection is derivable only from the measurement of distances between the query and the reference objects, the definition of the regions, and the search radius. Here for example the possible solution area shown is determined using only the four distance calculations $d(q, p_1)..d(q, p_4)$.

Table 1 shows how the example data objects a to f are stored in terms of their regional containment. The row labelled Q shows the containment relation between the entire query ball and each defined region. Where the boundaries intersect, a $\cap$ is shown; where they do not, a Boolean value shows whether the query ball is contained or otherwise. Therefore, the only possible solutions to the query are those which match on all non-intersecting fields; in this case, the objects a, e and f. Note that this is equivalent to the Conjunctive Normal Form (CNF) expression $B \wedge \neg C \wedge D \wedge \neg F$. This expression therefore covers the set of all possible solutions to the query.

The full set of query solutions can therefore be evaluated in three phases as follows: (a) the query is checked against the region definitions; (b) the useful information from this phase is used to conduct a series of bitwise operations to identify a set of candidate solutions for the query; and (c) the candidates are checked against the original set. This gives interesting performance tradeoffs; phase (b) should be almost constant cost, as a set of $\log n$ orthogonal, balanced partitions should exclude almost all incorrect solutions from the candidate set. The cost of phase (c) directly depends on how well this can be achieved, and will always be better with a larger number of regions which will require an increase in the cost of phase (a).

As dimensionality increases, then so will the proportion of intersections between the query and region boundaries. This can be countered by defining more regions, which will increase the storage cost of the index structure, and the cost of computing phase (a) of the query, but need *not* consequentially require any increase in memory or time for the ensuing phases of the computation.

2.2   Contribution

The contribution of this paper lies in the combination of using a fixed set of reference objects to characterise the data, followed by their use in an exact metric search algorithm which maximises the efficacy of memory use. There are many mechanisms which use similar fixed sets of reference objects (see for example Sect. 5); our mechanism is particularly well-suited to exact search as the dimensionality of the underlying data increases, by minimising the memory footprint required for the search algorithm. Furthermore our algorithm is inherently decomposable and parallelisable, and is well-suited to implementation on modern processors, including GPUs.

In this paper we give the basic mechanism, show its feasibility with respect to some well-known (relatively small and low-dimensional) data sets, and also show its performance as intrinsic dimensionality increases. This is an early exposition of these ideas, and there are many more aspects to study.

3   Core Mechanism

3.1   Data Structures

Before describing the algorithm in more detail, we describe the data structures used in the algorithm and their initialisation. We refer to a finite space (S, d) which is a subset of an infinite metric space (U, d).

A set P of enumerated reference objects p_0 to p_m is first selected from the finite metric space S. Based on this, we define a set of surfaces within U, defined according to the distance function d, each of which divide U into two parts. Surfaces within U are either *balls*, for example $\{u \in U \mid d(u, p_i) = \mu\}$ for some values i, μ, or *sheets*, for example $\{u \in U \mid d(u, p_i)) = d(u, p_j)\}$ for some values i, j. For each such surface, it is easy to categorise any element of u_i of U as being inside or outside an associated region, according to whether it is on the same side of the surface as p_i or otherwise. Note that there are many more regions than reference objects; for example a set of m reference objects immediately defines $\binom{m}{2}$ hyperplane regions, and can be used to define many more than this. In Sect. 3.3 we discuss further the selection of regions from the available reference objects.

We now define the notion of an *exclusion zone* as a containment map of S based on a given region; this is the information we will use at query time to perform exclusions and derive a candidate set of solutions. We impose an ordering on S, then for each s_i map whether it is a member of the region or otherwise. This logical containment information is best stored in a bitmap of n bits, where $n = |S|$. One such exclusion zone is generated per region and stored as the primary representation of the data set.

It is worth noting that an essential difference between our mechanism and others that use the same characterisation of the data (see Sect. 5) is that each of our bitmaps represents the containment of the whole data set within an individual region, rather than those regions which contain an individual object. The same information is thus divided with the opposite orientation; with reference to Table 1, we store the columns rather than the rows.

3.2   Query

The query process comprises three distinct phases as mentioned above:

Phase 1. Initially, the distance from the query q to each reference object p_i is measured. For each region, it can be established if the boundary of the solution ball intersects with the boundary of the region. For a ball region defined by reference object p_i and a radius μ, then the condition for intersection is

$$|d(p_i, q) - \mu| \leq t$$

For a sheet region, the condition depends on whether the metric d has the supermetric property (see Sect. 5) or otherwise: if it does, the intersection condition is
$$\frac{|d(p_i, q)^2 - d(p_j, q)^2|}{2d(p_i, p_j)} < t$$

otherwise the condition is
$$\frac{|d(p_i, q) - d(p_j, q)|}{2} < t$$

If the intersection condition holds, then the exclusion zone related to the region is not considered further; if it does not, then the exclusion zone is brought into the query calculation in one of two sets, depending on whether the query solutions are fully contained within, or without, the region in question. We will name these two sets of bitmaps B_{in} and B_{out}.

Phase 2. The second phase comprises the manipulation of the bitmaps deriving from the first phase to identify a set of candidate solutions. This may be efficiently achieved by a series of bitwise operations over these bitmaps. The solution to the query is guaranteed to lie within the intersection of the inclusion sets (derived by bitwise and operations) and not in the union of the exclusion set (derived by bitwise or). Thus any solution is guaranteed to be identified by the bitmap deriving from the following logical expression:

$$\left(\bigwedge_{b \in B_{in}} b \right) \wedge \left(\neg \left(\bigvee_{b \in B_{out}} b \right) \right)$$

Phase 3. The last phase consists of filtering the result sets derived in phase 2 against the original space and distance metric in order to produce an exact solution to the query.

3.3   Efficiency Considerations

No data is considered during Phase 1, and n distance calculations can be used to generate a great many judgements. With increasing dimensionality, a larger number of reference objects will be required to usefully characterise the space, as more queries will intersect with each region, and therefore a greater number of regions is required to maintain the size of $B_{in} \cup B_{out}$. However this adds no further cost to the second and third phases of the query.

If each bitmap used in the Phase 2 calculation is balanced, i.e. it contains the same number of 0s and 1s, and orthogonal, i.e. there is no logical dependency among them, then only $\log_2 n$ bitmaps (where $n = |S|$) are required to exclude almost all non-relevant data. As each bitmap is n bits long, this gives a space requirement of $O(n \log n)$, and a time requirement of $O(\log n)$. In this sense, the solution can not be said to be scalable. However it is important to look more deeply than this: the space requirement is literally $n \log_2 n$ bits, which cannot cause any real problem for any context where any n objects are being stored - even if n is huge, $\log_2 n$ bits is unlikely to approach the size of a single data object. If the bitmaps are huge, as there is no logical internal dependency they can be partitioned and accessed in parallel. Furthermore the time requirement on modern hardware is likely to approximate to a small constant time even with relatively large values of $log_2 n$.

Phase 3 is essentially optional, and required only for exact search. Alternatively, the first two phases can be considered as an approximate search technique. Whether this is desirable or not depends on how well the data is characterised according to the selected regions; in Sect. 4 we show an example where the number of false positives is so small this phase is hardly required.

Finally it can be noticed that each of the three phases of the computation are inherently parallelisable, and in particular Phase 2 should be extremely efficient on modern hardware, comprising as it does only parallelisable bitwise operations.

Balancing. In our initial experiments, we have tried both balanced and unbalanced bitsets. Balancing can be achieved by selecting a set of *witness* objects from the finite space S and finding a median distance or offset for these, so that the regional boundary divides the finite set into two equal parts. A large enough set of witness objects will give a good statistical approximation to the distribution of S. For ball partitions, the median distance to the centre is used; for sheet partitions, an offset can be selected left or right of the central hyperplane [5,11]. Furthermore, for supermetric spaces, the XY plane can also be rotated to maximise the spread of values as described in [8].

We still have much to investigate in terms of finding the best parameters for a given data set. In the meantime we note that balancing does increase performance with lower-dimensional spaces and smaller query thresholds, but that it starts to have a detrimental effect as either of these increases - the tradeoff being that balancing will increase the effectiveness of the second phase algorithm, but decrease the effectiveness of the first phase. In general, as query radius or dimensionality increases, we perceive that this is best offset by an increase in reference objects but also some controlled rebalancing of regional offsets.

4   Experiments and Results

Experiment 1. In the first experiment we investigate the efficacy of the algorithm by running queries against the SISAP *nasa* and *colors* data sets [9]. The metric used for both datasets is Euclidean distance. The number of reference objects is varied from 10 to 60 in steps of 5 and queries comprising 10% of the dataset are made against the remaining 90%. Thresholds of $t_0 = 0.12, t_1 = 0.285$, and $t_2 = 0.53$ are used for *nasa* and $t_0 = 0.052, t_1 = 0.083, t_2 = 0.13$ for *colors*. In this set of experiments the number of ball radii is set to 5 with the radii being set to a mean radius of 1.81.[1] and mean ± 0.3 and mean ± 0.6. We report residual distance calculations, which are the number of calculations made in phase 3 of the algorithm, excluding reference object distance calculations. The results of this experiment are shown in Fig. 2, with the figures for 60 reference objects (the right hand side of the graph) given in Table 2. To put these figures into the metric indexing context, the top two rows of the table give the number of distance calculations per query reported in [8] for the Distal SAT operating with both normal metric and supermetric exclusion mechanisms.

The three sets of graphs shown in Fig. 2 illustrate a number of interesting facets of the algorithm. As would be expected, increasing the number of reference objects has a significant effect of the number of distance calculations performed. However the number of reference objects used in these experiments is really quite small, much smaller than has been reported for other regional approaches e.g. [1,13] given the accuracy shown by the relatively small number of residual distance calculations required. In particular we have the quite stunning result that, using 60 reference objects, the supermetric property, and balancing the data structures, we have an almost perfect characterisation of the *nasa* data set, with less that one false positive result per query.

Experiment 2. The second experiment is designed to show how the algorithm can combat increasing dimensionality by using increased numbers of reference

---

[1] For balanced versions (see Sect. 3.3) the central radius is reset, but the same increments are applied.

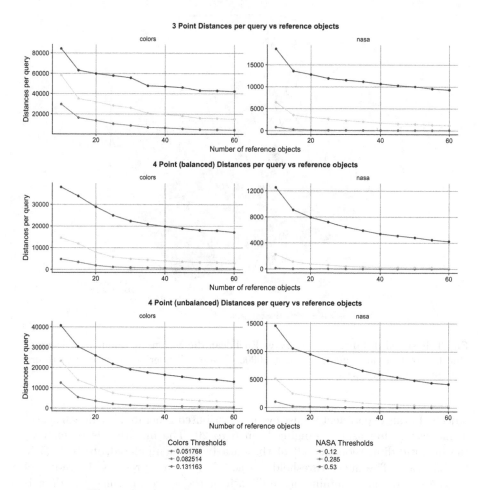

Fig. 2. Queries against SISAP colors and NASA datasets

Table 2. Residual distance calculations required when 60 reference objects are used (the numbers reported do not include the 60 distance calculations required for these.) The top two rows give comparable figures for the state-of-the-art Distal SAT.

	Colors			Nasa		
	t_0	t_1	t_2	t_0	t_1	t_2
DiSAT: metric	4049	9112	19745	554	2176	6448
DiSAT: supermetric	2015	5737	16199	320	1300	5444
Metric, unbalanced	4207	14930	42139	14	1120	9202
Metric, balanced	2246	9610	30250	11	822	7671
Supermetric, unbalanced	544	**3114**	**13045**	3	296	4122
Supermetric, balanced	**518**	3259	17512	1	**91**	**2204**

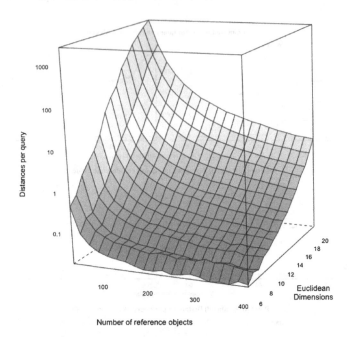

Fig. 3. Residual distances per query vs Euclidean dimensions vs number of reference objects: 1000 random queries against 10000 objects; all queries callibrated to return one-millionth of the space

objects. In this experiment we use evenly-distributed generated Euclidean spaces of increasing dimensions, ranging from 6 to 20. The intention is to increase the intrinsic dimensionality [4] of the data being manipulated, up to 27.5 in the last case. The query threshold at each dimension represents the radius of a hypersphere that contains one-millionth of the space in which the data is generated[2]. In all cases the number of points in the space is ten thousand and one thousand random queries are made against this data set. The number of reference objects ranges from 20 to 400. Results are shown in Fig. 3.

The surface shown in the image reports the number of residual distance calculations required per query plotted on a logarithmic scale, as both dimensions and number of reference points increase. Notice that the mid-point of the Z-axis (the number of residual calculations) represents accessing only 0.1% of the data, thus the figure demonstrates highly tractable search all the way to 20 dimensions. We do not believe that this has been demonstrated previously for any exact search mechanism.

Furthermore it can be seen that much of the surface has a z-value of substantially less than 1; in these cases, the space is effectively being perfectly characterised by the regions derived from the reference points. For example this

---

[2] For dimension n, radius $r_n = \dfrac{\Gamma(\frac{n}{2}+1)}{\pi^{\frac{n}{2}}}$, where Γ is Euler's gamma function.

occurs with less than 200 reference points for 10 dimensions, a space normally considered to be at the bounds of tractability for most exact search mechanisms.

5   Related Work

5.1   Search Using Fixed Reference Objects

There are a number of well-known mechanisms in which the distances between a query and a fixed set of reference objects are used to guide a first phase of search. All such mechanisms may either be regarded as approximate, or subsequently checked against the original data set for accuracy.

LAESA [14] has typically been used for metric filtering, rather than approximate search. For each element of the data, distances to a fixed set of reference points are recorded in a table. At query time, the distances between the query and each reference point are calculated; the table can then be scanned row at a time, and each distance compared; if, for any reference object p_i and data object s_j the absolute difference $|d(q, p_i) - d(s_j, p_i)| > t$, then it is impossible for s_j to be within distance t of the query, and the distance calculation can be avoided. LAESA can be used as an efficient pre-filter for exact search when memory size is limited. The same data can be re-cast into a metric space using the Chebyshev metric but this does not typically result in any significant performance increase.

The best known mechanisms which use a fixed partition of the original data for approximate search are based on *permutation orderings* [1,10,15]. There is a significant variety of techniques, but the essence of the approach is to characterise each element of the data in terms of its distances to a set of pre-selected reference objects, and then to compare elements using various aspects of similarity in this ordering. The approach is similar in that this effectively creates a large number of regions within the universal space, one for each ordering. However in all cases that we know of, an approximate search mechanism is created based on some cost function over the resulting orderings; there does not seem to be any clear mechanism for using the regions thus defined as an exact search mechanism. Especially in higher dimensional spaces, two very similar objects in the original space may lie on opposing sides of a number of the important boundaries, which then appear as distant objects and fail to appear in the search results.

The approach with *Sketches* [13] is more like our own, and the characterisation of the data as a set of bitmaps is almost identical. The approach differs however as it proceeds to use each bitmap as an object proxy with other search techniques, the underlying notion being the probabilistic mechanism that two close objects in the original space will result in two very similar bitmaps; thus the Hamming distance over these bitmaps should give a good proxy to the distance in the original space. The technique suffers from the same probabilistic issues as permutation orderings, in that for any two very similar objects there is a finite probability that they will appear very different in the proxy space.

Other probabilistic mechanisms have also been proposed, for example [2,17, 18]. These are all quite similar in outline but with different ways of defining regional boundaries and treating the proxy space. Our work differs significantly in that we describe a mechanism which is guaranteed to give all correct results from the original space, and in the way that the increased cost of increasing dimensionality or query threshold can be controlled.

5.2   Metrics and Supermetrics

Much work on finite isometric embeddings was conducted in the 20th century, by e.g. Blumenthal [3], Wilson [19] and Menger [12]. Blumenthal uses the phrase *four-point property* to mean a space that is 4-embeddable in 3-dimensional Euclidean space: that is, that for any four objects in the original space it is possible to construct a distance-preserving tetrahedron.

This simple property has been shown to have profound consequences in metric search. Connor and Vadicamo have applied these results in theoretical mathematics to this more practical domain [6–8] and the term *supermetric* is now used to refer to metrics with the property. For this context, the important result is that the four-point property applies to many commonly-used distance metrics, including Euclidean, Cosine[3], Jensen-Shannon, Triangular and Quadratic Form distances, all of which can be safely used in conjunction with the mechanisms described here.

Use of the supermetric property gives better exclusion conditions, as detailed in Sect. 3.2. The concepts given in this paper are equally applicable to both metric and supermetric spaces; as would be expected from a space with tighter geometric properties, however the results we show using the four-point property are significantly better than results relying only on the three-point property.

6   Conclusions and Further Work

This paper presents a novel exact search mechanism which gives excellent performance compared with other exact methods. It is especially well suited to higher dimensional data, in cases where indexing methods become intractable. The first phase of the algorithm is evaluated independently of the data set, and its cost is approximately $\mathcal{O}(m^2)$ where m is the number of reference objects selected; however this cost has a low constant factor and the cost of the phase is dominated by the m distance calculations.

The actual selection computation requires as few as $\log_2 n$ bits per datum, in which context the space required, although proportional to $\mathcal{O}(n \log n)$, is unlikely to cause a practical problem. Even when n is 10^9 the computation still fits comfortably within the memory of a modern laptop computer.

The algorithm is inherently decomposable and parallelisable. Every phase of the computation is parallelisable: the distances between queries and reference objects, the assessment of queries against regions, the bitwise operations over the data representation, and finally the filtering of the candidate results. Amdhal's law applies to the end to end pipeline but we believe that great speedups are

---

[3] For the correct formulation, see [6].

possible by exploiting parallel hardware resources due to the nature of the algorithm. We have begun experimenting with GPUs to perform the bit operations but it is premature to report any results at this time.

The algorithm exhibits many opportunities for tuning which are yet to be explored. We have already demonstrated the effect of changing the number of reference objects and how this may help in combating increasing dimensionality. However it may be possible to employ techniques to choose pivots which could enable queries to perform a number of distance calculations that are closer to the theoretical minimum. We have already experimented with choosing pivots to optimise sheet exclusions but reporting on this work here would also be premature. The size, number and uniformity of the radii used for ball exclusions is also worthy of exploration.

Acknowledgements. This work was supported by ESRC grant ES/L007487/1 "Administrative Data Research Centre—Scotland". We would like to thank Tom Dalton for his help with preparation of the data and creating R scripts for rendering results, and Peter Christen along with the anonymous reviewers for helpful comments on earlier drafts.

References

1. Amato, G., Gennaro, C., Savino, P.: MI-file: using inverted files for scalable approximate similarity search. Multimed. Tools Appl. **71**(3), 1333–1362 (2014)
2. Andrade, J.M., Astudillo, C.A., Paredes, R.: Metric space searching based on random bisectors and binary fingerprints. In: Traina, A.J.M., Traina, C., Cordeiro, R.L.F. (eds.) SISAP 2014. LNCS, vol. 8821, pp. 50–57. Springer, Cham (2014). https://doi.org/10.1007/978-3-319-11988-5_5
3. Blumenthal, L.M.: A note on the four-point property. Bull. Am. Math. Soc. **39**(6), 423–426 (1933)
4. Chávez, E., Navarro, G., Baeza-Yates, R., Marroquín, J.L.: Searching in metric spaces. ACM Comput. Surv. **33**(3), 273–321 (2001)
5. Connor, R.: Reference point hyperplane trees. In: Amsaleg, L., Houle, M.E., Schubert, E. (eds.) SISAP 2016. LNCS, vol. 9939, pp. 65–78. Springer, Cham (2016). https://doi.org/10.1007/978-3-319-46759-7_5
6. Connor, R., Cardillo, F.A., Vadicamo, L., Rabitti, F.: Hilbert exclusion: improved metric search through finite isometric embeddings. ACM Trans. Inf. Syst. **35**(3), 17:1–17:27 (2016)
7. Connor, R., Vadicamo, L., Cardillo, F.A., Rabitti, F.: Supermetric search with the four-point property. In: Amsaleg, L., Houle, M.E., Schubert, E. (eds.) SISAP 2016. LNCS, vol. 9939, pp. 51–64. Springer, Cham (2016). https://doi.org/10.1007/978-3-319-46759-7_4
8. Connor, R., Vadicamo, L., Cardillo, F.A., Rabitti, F.: Supermetric search. Inf. Syst. (2018). https://doi.org/10.1016/j.is.2018.01.002
9. Figueroa, K., Navarro, G., Chávez, E.: Metric spaces library (2007). http://www.sisap.org
10. Chavez Gonzalez, E., Figueroa, K., Navarro, G., Navarro, G.: Effective proximity retrieval by ordering permutations. IEEE Trans. Pattern Anal. Mach. Intell. **30**(9), 1647–1658 (2008)

11. Lokoč, J., Skopal, Y.: On applications of parameterized hyperplane partitioning. In: Proceedings of the Third International Conference on SImilarity Search and APplications, SISAP 2010, pp. 131–132. ACM, New York (2010)
12. Menger, K.: New foundation of Euclidean geometry. Am. J. Math. **53**(4), 721–745 (1931)
13. Mic, V., Novak, D., Zezula, P.: Improving sketches for similarity search. Proc. MEMICS **2015**, 45–57 (2015)
14. Micó, M.L., Oncina, J., Vidal, E.: A new version of the nearest-neighbour approximating and eliminating search algorithm (AESA) with linear preprocessing time and memory requirements. Pattern Recogn. Lett. **15**(1), 9–17 (1994)
15. Mohamed, H., Marchand-Maillet, S.: Quantized ranking for permutation-based indexing. Inf. Syst. **52**, 163–175 (2015). Special Issue on Selected Papers from SISAP 2013
16. Rivero, L.C., Doorn, J.H., Ferraggine, V.E. (eds.): Encyclopedia of Database Technologies and Applications. Idea Group, Hershey (2005)
17. Silva, E., Teixeira, T., Teodoro, G., Valle, E.: Large-scale distributed locality-sensitive hashing for general metric data. In: Traina, A.J.M., Traina, C., Cordeiro, R.L.F. (eds.) SISAP 2014. LNCS, vol. 8821, pp. 82–93. Springer, Cham (2014). https://doi.org/10.1007/978-3-319-11988-5_8
18. Tellez, E.S., Chavez, E.: On locality sensitive hashing in metric spaces. In: Proceedings of the Third International Conference on SImilarity Search and APplications, SISAP 2010, pp. 67–74. ACM, New York (2010)
19. Wilson, W.A.: A relation between metric and Euclidean spaces. Am. J. Math. **54**(3), 505–517 (1932)
20. Zezula, P., Amato, G., Dohnal, V., Batko, M.: Similarity search - the metric space approach. In: Advances in Database Systems (2006)

Relative Minimum Distance Between Projected Bags for Improved Multiple Instance Classification

José Francisco Ruiz-Muñoz[1,2], Germán Castellanos-Dominguez[1], and Mauricio Orozco-Alzate[1(✉)]

[1] Universidad Nacional de Colombia - Sede Manizales, Manizales, Colombia
{jfruizmu,cgcastellanosd,morozcoa}@unal.edu.co
[2] Instituto Tecnológico Metropolitano, Medellín, Colombia

Abstract. A novel *relative minimum distance* is introduced that allows improving the dissimilarity-based multiple instance classification. To this end, we apply a previously proposed mapping that brings closer, at least, a single instance from each positive training bag, while the negative-bags instances are driven apart. Our results show an increased classification performance on a broad type of real-world datasets.

Keywords: Multiple Instance Learning · Metric Learning
Nearest-neighbor rule

1 Introduction

Multiple Instance Classification (MIC) systems learn from patterns represented by sets of feature vectors (termed *bags*), where each bag holds multiple-feature objects (*instances*), so that a single label is attached to each bag, but not to every instance. Moreover, inside one bag, there might be instances which are either more related to other classes of bags or do not convey any information of its class membership, providing confusing information. To deal with this ambiguous representation, the binary classification setting separates the instances into *positive* and *negative*, defining the set formed by positive instances as the *concept*. Therefore, a bag is labeled as positive if and only if, at least, one of its instances belongs to the concept. Otherwise, the bag is negative. As a rule, the instance-level analysis implies the use of a classifier that assigns a single label to every single instance. However, this classification method neglects the extraction of more global structures that can reveal the interrelations between instances inside a bag, restraining a further improvement in the classification performance [1].

Intending to increase the performance of instance-level MIC algorithms, some characterization approaches have been proposed before that employ different pairwise dissimilarity measures for direct comparison between bags. More specifically, a single label can be inferred for each query bag to be represented by a

© Springer Nature Switzerland AG 2018
S. Marchand-Maillet et al. (Eds.): SISAP 2018, LNCS 11223, pp. 47–56, 2018.
https://doi.org/10.1007/978-3-030-02224-2_4

vector that holds the dissimilarities between bags and prototypes, resulting in a dissimilarity representation that is equivalent to a traditional feature vector space and thus a standard supervised classifier can be employed. In a first approach, *Bag-to-instance distances* can be devised for capturing the instance-level information that maps each bag into a high-dimensional feature vector like the Multiple Instance Learning via Embedded Instance Selection (MILES) in [2]. However, the performance of some classifiers (those that are affected by the curse of dimensionality) is limited since the number of prototype instances is often much higher than the number of bags [3]. In another approach, *Bag-to-bag distances*, which do not necessarily satisfy all the metric properties, allow working in low-dimensional spaces, but they tend to miss the instance-level information, i.e., the non-discriminant instances hide the discriminant ones. Therefore, the distance function must uncover the inherent *concept* to be learned, but performing well in high-dimensional representations at the same time. To this end, the dissimilarity measure must build a low-dimensional representation, and enhance the instance representation that most likely belongs to the concept, intending to benefit from the instance-level information.

For increasing the performance of the dissimilarity-based MIC algorithms, we propose the *relative minimum distance* that compares bags and prototype bags, relying on the information extracted from the mapped instances. We compute this mapping through the Multiple Instance Logistic Discriminant Metric Learning (MildML) algorithm proposed in [4]. Validation on several real-world datasets proves that the proposed dissimilarity measure between projected bags outperforms other baseline dissimilarities commonly accepted for MIC implementation.

2   Methods

2.1   Multiple Instance Classification

The MIC approach is a weakly supervised task of predicting class labels that represents each object in a feature space by a set of points $\mathcal{S} = \{s_1, \ldots, s_N\}$ (termed *bag of instances*) where $s_i \in \mathbb{R}^{n \times 1}$ is the i-th instance [5]. Each bag consists of two subsets of instances, one positive and another one negative: $\mathcal{S} = \overset{\oplus}{\mathcal{S}} \cup \overset{\ominus}{\mathcal{S}}$. Thus, a positive subset holds only positive instances ($\overset{\oplus}{\mathcal{S}} = \{s_1^+, \ldots, s_{N^+}^+\}$, with $N^+ \leq N$), i.e., the instances that are directly related to the *concept*. In contrast, a negative subset holds only the negative instances that are not related to the concept ($\overset{\ominus}{\mathcal{S}} = \{s_1^-, \ldots, s_{N^-}^-\}$, with $N^- \leq N$). Besides, $\mathcal{S}$ is positive if there is, at least, one instance in the bag relating to the concept ($\mathcal{S}^+$), that is, if $\overset{\oplus}{\mathcal{S}} \neq \varnothing$. Otherwise, $\mathcal{S}$ becomes negative ($\mathcal{S}^-$).

To find the instances in the training set that most likely belong to the concept, we apply the strategy of collapsing the class representation, as introduced in [6], resulting in learning a metric that produces low values when comparing objects of the same class or high values when comparing objects of different classes.

Therefore, the input space is mapped so that, at least, one instance of each positive training bag gets as close as possible to the others, driving apart the nearby instances of negative bags. To this end, we use the *MildML* method that maximizes the following loss function [4]:

$$\max_{L,b} \mathcal{L} = \max_{L,b} \sum_{k,l} \lambda_{kl} \log p_{kl} + (1 - \lambda_{kl}) \log(1 - p_{kl}) \tag{1}$$

where k and l are the indexes for the couple of compared bags, λ_{kl} is 1 if both $(\mathcal{S}_k$ and $\mathcal{S}_l)$ are positive bags. Otherwise, it is 0, $p_{kl} \in \mathbb{R}[0,1]$ is the probability that $\mathcal{S}_k$ and $\mathcal{S}_l$ are together positive, calculated by the *sigmoid* function as $p_{kl} = (1 + \exp(-b + d(\widetilde{\mathcal{S}}_k, \widetilde{\mathcal{S}}_l)^2))^{-1}$, where $b \in \mathbb{R}$ is a shifting parameter, $d(\cdot, \cdot) \in \mathbb{R}^+$ is a dissimilarity measure between the projected bags $\widetilde{\mathcal{S}} = \{\widetilde{s}_n^L : n \in N\}$, which hold the instances in the mapped space through a projection matrix $L \in \mathbb{R}^{n \times n}$. Note that the projection is a mapping $(\mathbb{R}^{n \times 1} \to \mathbb{R}^{n \times 1})$ applied to each instance.

The optimization task in Eq. (1) can be solved iteratively by the conjugate gradient algorithm [7], employing the update rule: $L_{k+1} = L_k + \alpha_k P_k$, where $\alpha_k \in \mathbb{R}$ is the step length, and $P_k \in \mathbb{R}^{n \times n}$ is the search direction that is fixed to $P_k = -G_k + \beta_k P_{k-1}$. In the case of the Polak-Ribiére update, the scalar-valued step $\beta_k \in \mathbb{R}$ is computed as $\beta_k = (\Delta \gamma_{k-1}^\top \gamma_k)/(\gamma_{k-1}^\top \gamma_{k-1})$, where $\Delta \gamma_k \in \mathbb{R}^{n^2 \times 1}$ denotes the vectorized version of the gradient $\Delta G_k = G_{k+1} - G_k$ (with $\Delta G_k \in \mathbb{R}^{n \times n}$).[1] Likewise, $\gamma_k \in \mathbb{R}^{n^2 \times 1}$ is the vectorized version of the loss function gradient $G_k = \partial \mathcal{L}/\partial L$ (with $G_k \in \mathbb{R}^{n \times n}$), defined for the MildML method in terms of L as $G_k = L \sum_{k,l} (\lambda_{kl} - p_{kl})(s_k - s_l)(s_k - s_l)^\top$. Consequently, the mapping shortens the distances between positive instances at the time as the distances between negative instances are extended, making the concept to be confined to a projected region that is narrower than in the original space.

2.2 Dissimilarity-Based Multiple Instance Classification

Provided a test bag $\mathcal{S}$ with Q prototype bags, the nearest-neighbor (1-NN) MIC predicts a label according to the class of the closest bag from the representation set $\{\mathcal{R}_1, \ldots, \mathcal{R}_Q\}$. Thus, the label assigned is the one that corresponds to the minimum entry of the dissimilarity vector $d_\mathcal{S} = [d(\mathcal{S}, \mathcal{R}_1) \ldots d(\mathcal{S}, \mathcal{R}_Q)]$; remember that d is not required to be a metric distance. For implementation of $d(\cdot, \cdot) \in \mathbb{R}^+$, we consider the following three symmetric dissimilarity measures between bags that have been previously applied in particular MIL problems [8]: (*i*) Hausdorff distance (d_H), (*ii*) Mean-minimum distance ($d_{\overline{\min}}$), and (*iii*) Overall-minimum distance ($d_{\min}$), respectively, computed as follows:

---

[1] As indicated in [4], the complexity to estimate ΔG_k, that is $\mathcal{O}(M(M + n)n)$, where M is the total number of instances in the training set and n is the dimension, can be reduced to $\mathcal{O}(M(M + n)n')$, by applying dimension reduction, such that $n' < n$.

$$d_{\mathrm{H}}(\mathcal{S}_k, \mathcal{S}_l) = \max(\max_q(\min_m d(\boldsymbol{s}_m^{(k)}, \boldsymbol{s}_q^{(l)})), \max_m(\min_q d(\boldsymbol{s}_m^{(k)}, \boldsymbol{s}_q^{(l)}))),$$

$$d_{\overline{\min}}(\mathcal{S}_k, \mathcal{S}_l) = \boldsymbol{E}\{\boldsymbol{E}\{\min_m d(\boldsymbol{s}_m^{(k)}, \boldsymbol{s}_q^{(l)}) : \forall q\}, \boldsymbol{E}\{\min_q d(\boldsymbol{s}_m^{(k)}, \boldsymbol{s}_q^{(l)}) : \forall m\}\},$$

$$d_{\min}(\mathcal{S}_k, \mathcal{S}_l) = \min_{m,q} d(\boldsymbol{s}_m^{(k)}, \boldsymbol{s}_q^{(l)}),$$

where $\mathcal{S}_k = \{\boldsymbol{s}_1^{(k)}, \ldots, \boldsymbol{s}_{N_k}^{(k)}\}$ and $\mathcal{S}_l = \{\boldsymbol{s}_1^{(l)}, \ldots, \boldsymbol{s}_{N_l}^{(l)}\}$ are the bags of instances that respectively hold N_k and N_l n-dimensional instances, $d(\boldsymbol{s}_m^{(k)}, \boldsymbol{s}_q^{(l)})$ is a dissimilarity measure between instances (we choose the Euclidean distance), and notation $\boldsymbol{E}\{\cdot\}$ stands for the expectation operator. Note that the 1-NN rule demands the following asymptotic behavior of dissimilarities between bags: $d(\mathcal{S}^+, \mathcal{R}^+) \approx 0$, $d(\mathcal{S}^+, \mathcal{R}^-) \gg 0$, $d(\mathcal{S}^-, \mathcal{R}^+) \gg 0$, and $d(\mathcal{S}^-, \mathcal{R}^-) \approx 0$.

2.3   A Relative Minimum Distance in a Projected Space

Since the negative instances can be present in either case of bags, positive or negative, the listed distances in Sect. 2.2 are not always appropriate, e.g., when the negative instances for all bags have been generated from the same Gaussian distribution, having small covariance. A more useful distribution of the instances is obtained by previously applying MildML: it induces that the distance between instances fulfills the following conditions: $d(\boldsymbol{s}_k^+, \boldsymbol{s}_l^+) \approx 0$, $d(\boldsymbol{s}_m^+, \boldsymbol{s}_q^-) \gg 0$, and $d(\boldsymbol{s}_m^-, \boldsymbol{s}_q^-) \gg 0$. However, the considered bag-to-bag distances fail in the following cases: (1) $d_{\mathrm{H}}(\mathcal{S}^+, \mathcal{R}^+) \gg 0$ or $d_{\mathrm{H}}(\mathcal{S}^-, \mathcal{R}^-) \gg 0$, if there is at least an instance $\boldsymbol{s}^-$ such that $\min_r d(\boldsymbol{s}^-, \boldsymbol{r}) \gg 0$ ($\boldsymbol{s}^- \in \mathcal{S}$ and $\boldsymbol{r} \in \mathcal{R}$), or at least a $\boldsymbol{r}^-$ such that $\min_s d(\boldsymbol{s}, \boldsymbol{r}^-) \gg 0$; (2) $d_{\overline{\min}}(\mathcal{S}^+, \mathcal{R}^+) \gg 0$ or $d_{\overline{\min}}(\mathcal{S}^-, \mathcal{R}^-) \gg 0$, if there are several instances $\boldsymbol{s}^-$ such that $\min_r d(\boldsymbol{s}^-, \boldsymbol{r}) \gg 0$, or several $\boldsymbol{r}^-$ such that $\min_s d(\boldsymbol{s}, \boldsymbol{r}^-) \gg 0$; and, (3) $d_{\min}(\mathcal{S}^-, \mathcal{R}^-) \gg 0$ because $\min_{s,r} d(\boldsymbol{s}^-, \boldsymbol{r}^-) \gg 0$.

Aiming to avoid the above-mentioned drawbacks in the distances between bags, we propose the relative minimum distance that, provided a test bag $\mathcal{S}$ and either case of prototypes $\mathcal{R}_r^*$, is defined as follows:

$$d_{\mathrm{rel}}(\mathcal{S}, \mathcal{R}_r^*) = \begin{cases} \dfrac{\sum_{i=1}^{Q^+} d_{\min}(\mathcal{S}, \mathcal{R}_i^+)}{\sum_{i=1}^{Q^+} d_{\min}(\mathcal{R}_r^+, \mathcal{R}_i^+)}, & \text{if } \mathcal{R}_r^* = \mathcal{R}_r^+ \\[2ex] \dfrac{\sum_{i=1}^{Q^+} d_{\min}(\mathcal{R}_r^-, \mathcal{R}_i^+)}{\sum_{i=1}^{Q^+} d_{\min}(\mathcal{S}, \mathcal{R}_i^+)}, & \text{if } \mathcal{R}_r^* = \mathcal{R}_r^- \end{cases}$$

where Q^+ is the number of positive prototypes. So, a measured value $d_{\mathrm{rel}}(\mathcal{S}, \mathcal{R}_r^+) < 1$ implies that $\mathcal{S}$ becomes "more positive" than the r-th positive prototype. In contrast, $d_{\mathrm{rel}}(\mathcal{S}, \mathcal{R}_r^-) < 1$ yields a "less positive" $\mathcal{S}$ than the r-th negative prototype. Consequently, the lower the value of $d_{\mathrm{rel}}(\cdot, \cdot)$ – the higher the probability of belonging to the class of the corresponding prototype. Then, the problematic issue of dissimilarity measures between bags is avoided for the 1-NN rule.

3   Experiments

For appraising the effectiveness of our method, we estimate the F-score as the classification performance, using two estimation methods: a 1-NN leave-one-bag-out validation and a bootstrapping test. The first one consists in training the classifier with all the examples of the dataset except one that is left for testing, repeating this such that all the examples are used for testing once. For the second estimation method, we divide each dataset into two halves: one for training and another for testing, repeating this procedure 20 times. For the sake of comparison, three commonly used classifiers are contrasted: 1-NN, MILES (that is the state-of-the-art classifier), and linear SVM. In the last case, C parameter is tuned by choosing the one reaching the best performance for the training set among the values $\{10^{-3}, 10^{-2}, 10^{-1}, 1, 10, 10^2, 10^3\}$. In the experimental setup, validation is accomplished using the real-world multi-label datasets described in Table 1.

Table 1. Real-world datasets described by the number of positive bags ($\mathcal{S}^+$), number of negative bags ($\mathcal{S}^-$), number of features per instance (Dimension), total number of instances and minimum and maximum (min and max, respectively) number of instances per bag.

Application	Dataset	$\mathcal{S}^+$	$\mathcal{S}^-$	Dimension	Instances	Min	Max
Molecule activity	Musk 1	47	45	166	476	2	4
	Musk 2	39	63	166	65598	1	1044
	Mutagenesis1	125	63	7	10486	28	88
	Mutagenesis2	13	29	7	2132	26	86
Images	Fox	100	100	230	1302	2	13
	Tiger	100	100	230	1220	1	13
	Elephant	100	100	230	1391	2	13
Audio classification	Brown Creeper	197	351	38	10232	2	43
	Winter Wren	109	439	38	10232	2	43
	Pacific-slope Flycatcher	165	383	38	10232	2	43
	Red-breasted Nuthatch	82	466	38	10232	2	43
	Dark-eyed Junco	20	528	38	10232	2	43
	Olive-sided Flycatcher	90	458	38	10232	2	43
	Hermit Thrush	15	533	38	10232	2	43
	Chestnut-backed Chickadee	117	431	38	10232	2	43
	Varied Thrush	89	459	38	10232	2	43
	Hermit Warbler	63	485	38	10232	2	43
	Swainson's Thrush	79	469	38	10232	2	43
	Hammond's Flycatcher	103	445	38	10232	2	43
	Western Tanager	46	502	38	10232	2	43

The results of 1-NN leave-one-bag-out test are reported in Table 2. Note that, in the original space, $d_{\overline{\min}}$ exhibits the best performance for most of the datasets.

An advantage of $d_{\overline{\min}}$ is that it considers the overall distribution of the bag of instances but is not so sensitive to outliers as d_H or $d_{\min}$. However, $d_{\overline{\min}}$ fails for some of the classification problems, e.g., with the `Hermit Thrush` dataset of the audio application, for which the number of negative instances may be remarkably higher than the number of positive instances in the positive bags. Also, d_{rel} fails in most of the cases, since it is not guaranteed that the concept is located in a particular region of the original space. On the other hand, in the projected space, d_{rel} remarkably outperforms the rest of the compared measures for most of the tested datasets. Nonetheless, the distance $d_{\overline{\min}}$ exhibits the best performance in some cases. When $d_{\overline{\min}}$ outperforms others, the bags of instances might mainly hold positive instances.

Table 2. 1-NN leave-one-bag-out test (F-score performance) applying the four studied dissimilarity measures between bags in the original space and the projected space. Best performances are highlighted in boldface.

Dataset	Original space				Projected space			
	d_H	$d_{\overline{\min}}$	$d_{\min}$	d_{rel}	d_H	$d_{\overline{\min}}$	$d_{\min}$	d_{rel}
Musk1	86.60	88.89	86.32	72.38	86.54	91.26	91.26	**100**
Musk2	80.00	71.91	70.45	60.00	78.26	80.43	88.64	**100**
Mutagenesis1	86.31	**87.45**	81.25	78.60	84.77	**87.45**	80.93	79.72
Mutagenesis2	43.48	45.45	52.17	40.91	41.67	**61.54**	50	59.26
Fox	56.60	62.50	53.27	68.77	59.16	68.14	77.87	**98.49**
Tiger	70.71	76.04	73.10	64.31	66.03	80.89	87.72	**100**
Elephant	76.78	81.65	79.07	69.50	75.00	82.70	80.65	**100**
Brown Creeper	75.37	**80.20**	68.38	68.95	73.32	75.40	68.38	72.05
Winter Wren	87.23	91.32	76.36	79.41	88.59	**94.12**	78.93	83.94
Pacific-slope Flycatcher	74.32	**83.79**	74.38	69.85	66.86	76.58	78.91	81.94
Red-breasted Nuthatch	67.50	75.47	57.65	49.66	54.88	67.01	70.39	**100**
Dark-eyed Junco	25.53	37.50	20.00	26.67	24.00	40.82	30.30	**97.44**
Olive-sided Flycatcher	63.59	**82.61**	56.97	57.75	70.72	72.36	64.82	67.15
Hermit Thrush	7.14	25.00	11.11	14.29	18.18	20.69	26.32	**96.55**
Chestnut-backed Chickadee	66.14	65.35	55.07	48.52	60.56	70.54	77.42	**94.42**
Varied Thrush	95.40	**99.44**	87.18	73.59	96.00	96.63	83.96	98.88
Hermit Warbler	68.75	75.76	68.12	77.78	71.43	76.12	70.06	**93.94**
Swainson's Thrush	77.03	86.53	52.50	30.13	65.31	69.51	65.74	**92.62**
Hammond's Flycatcher	93.94	98.52	97.56	34.77	78.92	98.08	97.63	**100**
Western Tanager	54.95	85.06	49.44	47.15	57.14	78.43	55.56	**96.70**

As mentioned in Sect. 2.3, some bag-to-bag distances are prone to behave in a non-intuitive way, since they may produce high values between correctly labeled bags of the same class, and small values between bags that belong to different classes. To illustrate the overall behavior of each estimated measure, we plot the normalized histograms obtained from the considered dissimilarity

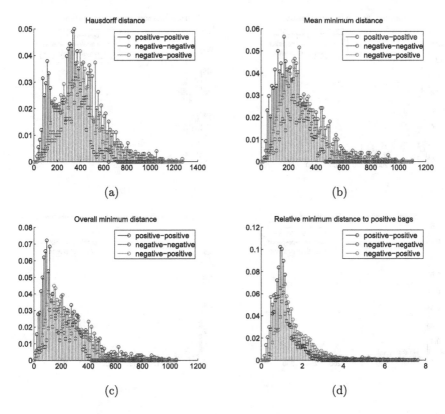

Fig. 1. Normalized histograms of distances in the original feature space (Musk1 dataset).

measures,—in the original space (see Fig. 1) as well as in the projected space (see Fig. 2)—between the following cases of bags in the Musk1 dataset: positive-positive bags, negative-negative bags, and negative-positive bags. Note that the histogram of a good dissimilarity measure should contain most of the distances between bags of the same class, positive-positive or negative-negative, on the left of the horizontal axis, and most of the distances between bags of different classes, negative-positive, on the right of the horizontal axis. Figure 1 shows that, in the original space, there is no clear difference in the distribution of the dissimilarities between neither bags of the same class, nor between bags of different classes. However, the mean of the dissimilarities calculated between positive-positive bags (for d_H, $d_{\overline{\min}}$ and $d_{\min}$) seems to be slightly smaller, than the dissimilarities between negative-negative and negative-positive bags. For d_{rel}, the dissimilarities between negative-negative bags are located on the left of those between positive-positive and negative-positive bags. On the other hand, Fig. 2 shows that applying the projection makes the distributions of the dissimilarities to behave in the above-mentioned intuitive way. Particularly, for $d_{\min}$, the dissimilarities between positive-positive bags are smaller, than those between

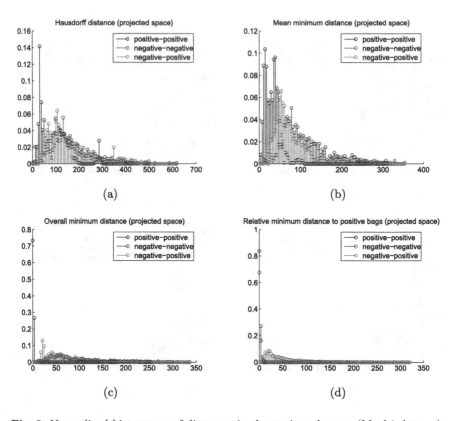

Fig. 2. Normalized histograms of distances in the projected space (Musk1 dataset).

negative-negative and negative-positive bags. For d_{rel}, the distances between bags of the same class are clearly lower than the ones between bags of different classes. It coincides with the outstanding results for this dataset reported in Table 2.

The performance using 1-NN and a linear SVM (applying d_{rel} in each case) against the state-of-the-art MILES classifier is reported in Table 3, showing that the linear SVM exhibits the best performance in most of the cases. Nonetheless, the performance achieved by the 1-NN classifier that strongly relies on the distance is still remarkable, if comparing to MILES method. Therefore, we prove that d_{rel} is appropriate to handle the sort of considered classification problems.

Table 3. Bootstrapping test (F-score performance) to compare the baseline MILES classification against the classification carried out by applying the proposed d_{rel} (1-NN d_{rel} and SVM d_{rel}). A dash indicates that classification of the target class completely failed.

Dataset	MILES	1-NN d_{rel}	SVM d_{rel}
Musk1	80.32 ± 7.84	**100**	**100**
Musk2	78.11 ± 7.06	**100**	99.40 ± 1.90
Mutagenesis1	$\mathbf{81.14 \pm 5.88}$	78.93 ± 3.04	80.55 ± 2.69
Mutagenesis2	$\mathbf{38.43 \pm 16.21}$	–	11.07 ± 18.49
Fox	63.56 ± 5.19	$\mathbf{99.60 \pm 0.80}$	98.89 ± 1.09
Tiger	77.14 ± 3.00	**100**	**100**
Elephant	75.25 ± 4.91	**100**	**100**
Brown Creeper	81.09 ± 2.18	84.86 ± 2.52	$\mathbf{91.73 \pm 1.82}$
Winter Wren	90.04 ± 2.65	85.13 ± 6.67	$\mathbf{95.47 \pm 2.11}$
Pacific-slope Flycatcher	79.69 ± 2.42	82.00 ± 7.95	$\mathbf{95.28 \pm 1.34}$
Red-breasted Nuthatch	82.12 ± 4.20	99.65 ± 0.56	**100**
Dark-eyed Junco	–	–	–
Olive-sided Flycatcher	74.69 ± 4.42	84.29 ± 14.47	$\mathbf{96.22 \pm 2.95}$
Hermit Thrush	–	–	$\mathbf{84.81 \pm 24.07}$
Chestnut-backed Chickadee	69.14 ± 5.32	76.34 ± 12.17	$\mathbf{95.83 \pm 2.09}$
Varied Thrush	99.00 ± 1.13	**100**	**100**
Hermit Warbler	71.81 ± 4.37	87.15 ± 8.08	$\mathbf{98.75 \pm 1.00}$
Swainson's Thrush	80.25 ± 4.56	87.61 ± 9.42	$\mathbf{93.98 \pm 2.19}$
Hammond's Flycatcher	98.89 ± 1.05	**100**	**100**
Western Tanager	84.20 ± 4.55	$\mathbf{90.77 \pm 20.30}$	88.5 ± 9.62

4   Concluding Remarks

We introduce the relative minimum distance for positive bags, denoted by d_{rel}, that globally compares a query bag to a prototype bag. This bag-to-bag distance is appropriate to build low-dimensional dissimilarity-based representations. Furthermore, we benefit from instance-level information, by previously computing a supervised projection, using MildML. In our experiments, we classify datasets of molecule activity, images, and audio. Our results show that applying d_{rel} in the projected space improves the classification performance in most of the studied datasets, in comparison with conventional distance measures between bags applied in either, the original or the projected space, and with a state-of-the-art multi-instance classifier. Our algorithm could fail if the projected space is not estimated as expected, e.g., if an insufficient number of iterations for MildML are computed or if increasing the distance between negative instances is not possible, because it is zero for many of them.

As future work, we propose a further study on the stopping criterion for MildML, on how to handle problems with many similar negative instances, and in reducing the complexity in the minimization of the loss-function to carry out the mapping. In addition, a throughout comparison against related methods such as citation k-NN [9] and the algorithm proposed in [10] is also a matter of further study.

Acknowledgment. This work is partially supported by "Convocatoria 567 de 2012 - Colciencias". The authors acknowledge support to attend SISAP 2018 provided by "Convocatoria para la Movilidad Internacional de la Universidad Nacional de Colombia (UNAL) 2017–2018".

References

1. Amores, J.: Multi-instance classification: review, taxonomy and comparative study. Artif. Intell. **201**, 81–105 (2013)
2. Chen, Y., Bi, J., Wang, J.Z.: MILES: multi-instance learning via embedded instance selection. IEEE Trans. Pattern Anal. Mach. Intell. **28**(12), 1931–1947 (2006)
3. Cheplygina, V., Tax, D.M.J., Loog, M.: Dissimilarity-based ensembles for multiple instance learning. IEEE Trans. Neural Netw. Learn. Syst. **27**(6), 1379–1391 (2016). https://doi.org/10.1109/TNNLS.2015.2424254
4. Guillaumin, M., Verbeek, J., Schmid, C.: Multiple instance metric learning from automatically labeled bags of faces. In: Daniilidis, K., Maragos, P., Paragios, N. (eds.) ECCV 2010. LNCS, vol. 6311, pp. 634–647. Springer, Heidelberg (2010). https://doi.org/10.1007/978-3-642-15549-9_46
5. Du, R., Wu, Q., He, X., Yang, J.: MIL-SKDE: multi-instance learning with supervised kernel density estimation. Sig. Process. **93**(6), 1471–1484 (2013)
6. Globerson, A., Roweis, S.T.: Metric learning by collapsing classes. In: NIPS, pp. 451–458 (2006)
7. Diene, O., Bhaya, A.: Conjugate gradient and steepest descent constant modulus algorithms applied to a blind adaptive array. Sig. Process. **90**(10), 2835–2841 (2010)
8. Cheplygina, V., Tax, D.M.J., Loog, M.: Multi-instance learning with bag dissimilarities. Pattern Recognit. **48**(1), 264–275 (2015)
9. Wang, J., Zucker, J.D.: Solving the multiple-instance problem: a lazy learning approach. In: Proceedings of the Seventeenth International Conference on Machine Learning, pp. 1119–1126. Morgan Kaufmann Publishers Inc. (2000)
10. Jin, R., Wang, S., Zhou, Z.H.: Learning a distance metric from multi-instance multi-label data. In: 2009 IEEE Conference on Computer Vision and Pattern Recognition, pp. 896–902 (2009)

Visual Search

Scalability of the NV-tree: Three Experiments

Laurent Amsaleg[1]([⊠]), Björn Þór Jónsson[2], and Herwig Lejsek[3]

[1] CNRS–IRISA, Rennes, France
laurent.amsaleg@irisa.fr
[2] IT University of Copenhagen, Copenhagen, Denmark
[3] Videntifier Technologies, Reykjavik, Iceland

Abstract. The NV-tree is a scalable approximate high-dimensional indexing method specifically designed for large-scale visual instance search. In this paper, we report on three experiments designed to evaluate the performance of the NV-tree. Two of these experiments embed standard benchmarks within collections of up to 28.5 billion features, representing the largest single-server collection ever reported in the literature. The results show that indeed the NV-tree performs very well for visual instance search applications over large-scale collections.

1 Introduction

Visual instance search (VIS) is the task of retrieving from a media collection the items that contain an actual instance of a visual query. Various real world applications require such fine-grained recognition capabilities, including forensics and copyright enforcement. A common theme in these applications is that the media collection is very large, calling for scalable indexing methods.

Due to the fine-grained nature of VIS, this domain will always require many local features for each media item. Query processing will then boil down to running multiple k-NN queries and consolidating the result into a single reply. Currently, SIFT features are the state-of-the-art for the VIS domain when considering extremely large image datasets: hundreds of local features are typically generated for each media item and result consolidation can be done via simple voting schemes. The scalability problem is compounded due to the multiple features generated: a VIS system handling tens of millions of images, for example, may need to manage tens of billions of local features.

At industry scale, VIS applications have the following requirements:

- The visual query typically results in multiple query features. Approximate indexing schemes are thus applicable to VIS applications, and in fact they can often tolerate fairly low recall.
- Each k-NN query is actually looking only for very specific features; in the case of copy detection applications typically only one feature. The appropriate quality metric is thus Recall@k.

S. Marchand-Maillet et al. (Eds.): SISAP 2018, LNCS 11223, pp. 59–72, 2018.
https://doi.org/10.1007/978-3-030-02224-2_5

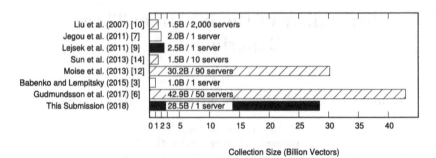

Fig. 1. The largest experimental collections reported in the literature. Black bars represent results using the NV-tree. Shaded bars represent multi-server configurations.

- Due to the scale of the feature collections, data may not fit in memory. The high-dimensional index must thus support disk-based processing.
- VIS applications are typically part of a larger pipeline, so query processing costs must be small and, no less importantly, predictable.

In previous work, we have proposed the NV-tree [8,9], a scalable approximate high-dimensional specifically designed for large-scale VIS applications. The NV-tree builds upon a combination of projections of data points to lines and partitioning of the projected space. By repeating the process of projecting and partitioning, data is separated into small neighborhoods which can be easily fetched from disk with a single read. By construction, the NV-tree thus guarantees query processing cost of at most one disk read per index per query feature.

Figure 1 shows a summary of all works reported in the literature with collections of at least 1B features. As the figure shows, the NV-tree has already been used for the largest reported single-server installation, with 2.5B features. The only two other systems that have been applied at this scale are Product Quantisation (PQ) [7] and Inverted Multi-Index (IMI) [2,3], while only installations based on Hadoop [12] and Spark [6] have used larger collections. The figure also shows that the largest collections reported in this paper contain 28.5B features; an order of magnitude more than the other single-server approaches have managed.

In this paper, we report on three experiments designed to evaluate the suitability of the NV-tree for VIS applications:

- In Sect. 3 we report on two image-based VIS benchmarks, embedded within collections of up to 28.5B distracting features, the largest single-server collection reported in the literature.
- In Sect. 4 we report on a single-query benchmark from our previous work, again embedded within collections of up to 28.5B features.
- In Sect. 5 we compare the NV-tree to PQ and IMI using the relatively small SIFT1B benchmark. While this is not a VIS benchmark, we use it because it

represents the largest collection used by those systems. We then analyse how PQ and IMI would perform with the 28.5B collection.

The results show that the NV-tree does very well for the large-scale VIS applications. With the small-scale non-VIS benchmark, the NV-tree has better performance but lower result quality. Further, our analysis shows that query processing performance of PQ and IMI would suffer with the 28.5B collection.

The NV-tree is proven technology, already in use at Videntifier Technologies, one of the main players in the forensics arena with technology deployed at such clients as Interpol. Their search engine targets fine-grained VIS for investigations that, e.g., aim to dismantle child abuse networks. The search engine can identify very fine-grained details in still images and videos from a collection of 150 thousand hours of video, typically scanning videos at 40x real-time speed, and while allowing dynamic insertions of about 700 h of video material every day.

2   The NV-tree

The NV-tree [8,9] is a disk-based high-dimensional index. It builds upon a combination of projections of data points to lines and partitioning of the projected space. By repeating the process of projecting and partitioning, data is separated into small partitions which can be easily fetched from disk with a single read, and which are likely to contain all the close neighbors in the collection. We briefly describe the NV-tree creation process, its search procedure, its dynamic insert process and then enumerate some salient properties of the NV-tree.

2.1   Index Creation

Overall, an NV-tree is a tree index consisting of: (a) a hierarchy of small *inner nodes*, which guide the feature search to the appropriate leaf node; and (b) larger *leaf nodes*, which contain references to actual features. The leaf nodes are further organised into *leaf-groups* that are disk I/O units, as described below.

When tree construction starts, all features from the collection are first projected onto a single projection line through the high-dimensional space ([8] discusses projection line selection strategies). The projected values are then partitioned in 4 to 8 partitions based on their position on the projection line. Information about the partitions, such as the partition borders along the projection line, forms the first inner node of the tree—the root of the tree. To build the subsequent levels of the NV-tree, this process of projecting and partitioning is repeated recursively for each and every partition, using a new projection line for each partition, thus creating the hierarchy of smaller and smaller partitions represented by the inner nodes.

At the upper levels of the tree, with large partitions, the partitioning strategy assigns equal distance between partition boundaries at each level of the tree. The partitioning strategy changes when the features in the partition fit within 6×6

leaf nodes of 4 KB each. In this case, all the features from that partition are partitioned into a *leaf-group* made of (up to) 6 inner nodes, each containing (up to) 6 leaves. In this leaf-group, partitioning is done according to an equal cardinality criterion (instead of an equal distance criterion). Finally, for each leaf node, projection along a final random line gives the order of the feature identifiers and the ordered identifiers are written to disk. It is important to note that the features themselves are *not* stored; only their identifiers.

Indexing a collection of high-dimensional features with an NV-tree thus creates a tree of nodes keeping track of information about projection lines and partition boundaries. All the branches of the tree end with leaf-groups with (up to) 36 leaf nodes, which in turn store the feature identifiers.

2.2   Nearest Neighbor Retrieval

During query processing, the search first traverses the hierarchy of inner nodes of the NV-tree. At each level of the tree, the query feature is projected to the projection line associated with the current node. The search is then directed to the sub-partition with center-point closest to the projection of the query feature until the search reaches a leaf-group, which is then fully fetched into RAM, possibly causing one single disk I/O. Within that leaf-group, the two nodes with center-point closest to the projection of the query feature are identified. The best two leaves from each of these two nodes are then scanned in order to form the final set of approximate nearest neighbors, with their rank depending on their proximity to the last projection of the query feature. The details of this process can be found in [9].

While the NV-tree is stored on disk, the hierarchy of inner nodes is read into memory once query processing starts, and remains fixed in memory. The larger leaf nodes, on the other hand, are read dynamically into memory as they are referenced. If the NV-tree fits into memory, the leaf nodes remain in memory and disk processing is avoided, but otherwise the buffer manager of the operating system may remove some leaf nodes from memory.

2.3   Properties of NV-trees

The experiments and analysis of [9] show that the NV-tree indexing scheme has the following properties:

- *Scalar Quantization:* The NV-tree uses random projections to turn multi-dimensional features into single-dimensional values indexed by B^+-trees. As only feature identifiers are stored, the NV-tree requires 6 bytes per feature.
- *Single Read Performance Guarantee:* In the NV-tree, leaf-groups have a fixed size. Therefore, the NV-tree guarantees query processing time of a single read regardless of the size of the feature collection. Larger collections need deeper NV-trees but intermediate nodes fit easily in memory and tree traversal cost is negligible.

– *Compact Data Structure:* The NV-tree stores in its index the identifiers of the features, not the features themselves. This amounts to about 6 bytes of storage per features on average. The NV-tree is thus a very compact data structure. Compactness is desirable as it maximizes the chances of fitting the tree in memory, thus avoiding disk operations.
– *Consolidated Result:* Random projections produce numerous false positives that can be almost all eliminated by an ensemble approach. Aggregating the results from a few NV-trees, which are built independently over the same collection, dramatically improve result quality.

3 Experiment 1: Image Benchmarks at Scale

In this first experiment, we have adopted a traditional fine-grained quasi-copy paradigm, implemented using SIFT features. We report on two benchmarks from the literature, embedded in collections of up to 28.5 billion "distracting" features. The resulting feature collection is so large (3.7 TB) that no other high-dimensional indexing scheme from the literature could handle it. The NV-tree could therefore not be compared to any other scheme from the literature. This applies to the results presented in this section, as well as the ones reported in Sect. 4.

3.1 Image Datasets and Ground Truth

The image retrieval benchmarks are the "49k" benchmark [1] and the "Copydays" benchmark [5]. These benchmarks apply predefined image transformations to a particular collection of images to obtain a set of query images. We then "drown" the original images, used to create the transformed quasi-copies, within a large collection of random images which play the role of "distracting" the search. The transformed query images are then evaluated against the indexed collection and the location of the original image in the final result list is noted. When the first image in the ranked result list is the original image, the answer is considered correct; if the first image in the ranked list is not the original image, then the system is said to fail, even if that image turns out to be second in the ranked list. For each transformation, 100% success means that all the ground truth images were at the top of the corresponding ranked lists in the result set.

All pictures used in our experiments were resized such that their longer edge is 512 pixels long. We did this to ensure that the number of SIFT features [11] computed over each image was about one thousand on average. All the original images used to create the quasi-copies of the 49k and Copydays benchmarks were resized accordingly. Queries and their counterparts are therefore consistent with respect to the distracting image collection within which they are drowned.

49k Image Benchmark. For this benchmark, one thousand images were randomly selected from Flickr, resulting in a very diverse collection. For each

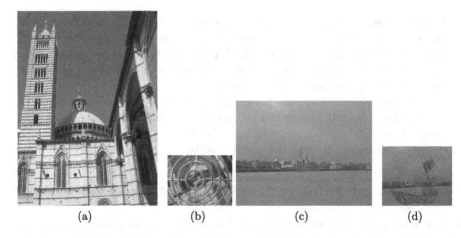

(a)   (b)   (c)   (d)

Fig. 2. Examples from Copydays: (a) and (c) are two original images, while (b) and (d) are two strong variants used as queries. Size ratio is preserved.

image, the Stirmark software [13] was used to generate 49 different transformations, summarized in [1]. Overall, this process generates 49,000 quasi-copy query images, hence the name "49k". Note that some of these quasi-copies are quite dissimilar from their original counterpart. For example, the CONV_2 transform tends to be extremely dark, to the point where very few SIFT features can be computed from the transformed images, making this quasi-copy very hard to find. The MEDIAN_9, NOISE_5 and PSNR_50 are also quite different, making the identification of their original counterparts challenging, especially because the SIFT features of the quasi-copies are either at different scales or at different locations in the images, since the visual noise produces very different local DoG extrema. Finally, it is worth noting that crops are inherently challenging since they dramatically reduce the number of SIFT features that can be computed for the quasi-copies, which in turn strongly decreases the number of possible matches. This is a truly difficult instance search problem.

Copydays Image Benchmark. This benchmark is a publicly available collection [5] used in several publications. We use this benchmark for our experiments because it contains quasi-copies that are more severely distorted than the ones in the 49k benchmark, making the original images much harder to find. Copydays contains 157 original images. Three families of transformations have been applied, as summarized in [5], resulting in 3,055 quasi-copies in total. Some of the 229 manual transformations are particularly difficult to find since they generate quasi-copies that are visually extremely different from their original counterparts. For example, Fig. 2 shows two original images along with one strong (manually created) variant.

Table 1. Distracting image collections used in experiment 1.

Collection	Images	SIFTs	Disk size
300M	334,268	305,443,749	40.3 GB
3B	2,970,596	3,040,856,472	401 GB
28.5B	28,969,271	28,484,904,924	3.7 TB

Distracting Image Collections. To evaluate the result quality produced by the NV-tree, we inserted the original images of the 49k and Copydays image sets into a larger set of images randomly downloaded from Flickr between 2009 and 2011. The downloading process rejected images smaller than 100×100 pixels and also used MD5 signatures to reject exact duplicates of any previously downloaded images. We have gathered almost 30 million such images, and we varied the number of distracting images in order to study the impact of collection size on the result quality of the indexing scheme. We report on experiments with three different data collections, all including the original images from the 49k and Copydays image sets, but having a different number of distracting images, resulting in roughly 300M, 3B and 28.5B features (see Table 1).

3.2   Result Quality

49k Image Benchmark. Figure 3 shows the result quality when querying three NV-trees with the 49 transformations, varying the distracting image collections. Figure 3 reads as follows: 100% of the ground truth images are found for 8.16% of the queries when the distracting collection is the 300M set. This means that all 1,000 ground truth images were ranked #1 in the result set for 4 of the 49 transformations. Continuing with the 300M distracting set, then 43 transformations are found from 90% to 100% of the time, or 87.17% of the queries.

Moving to larger distracting sets, 6.12% of the transformations are always found within the 3B set while this percentage goes down to 4.08% with the 28.5B set. In turn, more transformations are found between 90% and 100% of the times as depicted by the large gray area which grows to 91.84% with 3B and 93.88% with 28.5B. Overall, image retrieval works extremely well as almost all the ground truth images are found when queried with images belonging to the 49k image benchmark, with all three distracting collection sizes.

A few comments are in order, however. First, crops are always found, validating instance search. Second, one image variant is almost never found, no matter which distracting image collection is used; this is the "CONV_2" variant, which produces almost completely black images where next to no detail remains. It is common that the computation of the SIFT features on these images produces no features at all, or only one or two features, making the results random. Third, for some transformations, 100% recall is not reached. For example, for the collection that contains 28.5B features, "JPEG_15" gives 98.5%, "RML_10" gives 98.4% and "MEDIAN_9" gives 96.5%. A detailed analysis of the result lists shows that

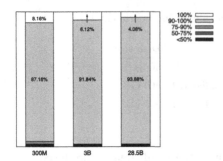
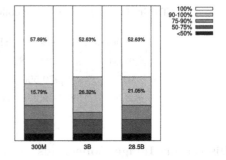

Fig. 3. Result quality for 49k.   **Fig. 4.** Result quality for Copydays.

in *all these cases*, the ground truth images are ranked from #2 (the most frequent situation) to #5. This is an extremely good result, especially because our
success criterion—considering only rank #1—is very strict.

Copydays Image Benchmark. Figure 4 shows the result quality for the Copydays image benchmark. Overall, the results are excellent for all but the most
difficult variants. The NV-tree is able to identify the correct images most of the
time, even from quite strongly distorted queries. It is not surprising to observe
that quality drops with extremely compressed images (a person can sometimes
hardly find any similarity between a JPEG 3% compressed image and its original version) and with some of the strong variants. Note that sometimes such
attacked query images create only a handful of features, so there are too few
matches for the original to rank #1—it is lost in the noise.

3.3   Retrieval Performance

We ran experiments using a Dell r710 machine that has two Intel X5650 2.67 GHz
CPUs. Each CPU has 12 MB of L3 cache that is shared by 6 actual and 6 virtual
cores. The RAM consists of 18×8 GB 800 MHz RDIMMs chips for a total of
144 GB. That machine is connected to a NAS 3070 storage system from NetApp,
offering about 100TB of magnetic disk space in a RAID configuration. We ran the
experiments using a single core, using three NV-trees which are probed one after
the other; no parallelism is enforced in our experiments to simlipfy interpretation
of the results.

When the three NV-trees fit entirely in main memory, which is the case for
the 300M and the 3B collections, answering each query feature is extremely
fast. A detailed analysis shows that, on average, 2,500 query features can be
processed per second. On average, therefore, identifying 100 near neighbors of a
single query feature takes about 0.4 ms per NV-tree. In turn, as there are about
1,000 query features per image, it takes about 400 ms to identify the images that
are the most similar to the query image.

When using the 28.5B collection, however, no index fits entirely in main memory. In this case, the system must therefore read data from disks for almost every query feature; as each query feature is likely to access a different part of the index, no main memory buffering policy is effective. Detailed analysis shows that about 50 query features could be processed per second in this case, which is 50 times slower than for the cases where the index fit in RAM. In the case of the 28.5B collection, it is thus possible to return the answer to a single query feature in 22.47 ms on average, while an image takes about 2.25 s.

Note that since some queries have very few features while others have more features, the retrieval time varies significantly. As pointed out earlier, however, the construction process of the NV-tree is such that the search requires only a single disk read. Because three NV-trees are used, no more than three disk reads are thus performed per feature search, and only 3,000 features are considered on average, out of the 28.5B, which is about 0.00001% of the collection.

4   Experiment 2: Single Feature Recall at Scale

This section again analyses the ability of the NV-tree index structure to cope with truly large-scale data collections, reporting results with up to 28.5B features, this time focusing on the recall of single query features.

4.1   Experimental Setup

In this experiment, we use the ground truth defined in [8]. A sequential scan was used to determine the 1,000 nearest neighbors of 500,000 query features, all coming from a collection of 180 million SIFT features. Analyzing the resulting 500M neighbors, we identified 248,212 features as being contrasted enough to be considered the true nearest neighbors of the query features. Contrast here is directly derived from criterion of [11]: a neighbor is considered a true neighbor if it is significantly closer than neighbor number one hundred.

We then embed these 248,212 features within feature sets of varying cardinalities to distract the search. These sets of distracting features have been created by extracting SIFT features from up to 30 million images randomly downloaded from Flickr. The images are ignored here, however, as we focus on individual query features. The resulting distractor sets are shown in Table 2.

This experiment focuses on recall, i.e., how many of these 248,212 ground truth features are found using the original set of 500,000 queries when varying the number of distractors, but we also report on retrieval performance. We are not aware of any other experiment ever published where recall measurements are obtained from searching the nearest neighbors of individual query features lost within 28.5B distracting features.

4.2   Quality of Nearest Neighbor Retrieval

Figure 5 shows the recall for the different distractor collections. The x-axis shows the size of the distractor collections (note the logarithmic scale), while the y-axis

Table 2. Distracting feature collections for experiment 2.

Collection	SIFT features	Feature collection size	NV-tree size
30M	28,799,690	3.8 GB	180 MB
180M	179,443,881	23.6 GB	1 GB
300M	305,443,749	40.3 GB	1.9 GB
2.5B	2,485,568,191	328 GB	14 GB
3B	3,040,856,472	401 GB	17 GB
28.5B	28,484,904,924	3.7 TB	162 GB

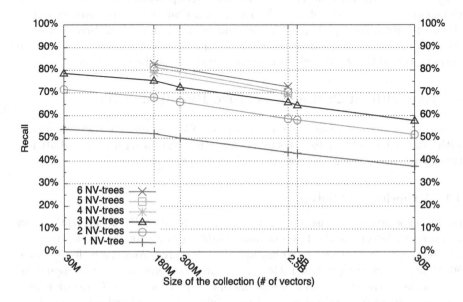

Fig. 5. Recall, varying collection sizes, varying number of NV-trees.

shows the impact on recall of using a varying number of NV-trees. Up to three
NV-trees were used against all the collections. We also considered using up to
six NV-trees to improve recall; as such experiments are complicated and time
consuming, however, we used only two moderate size datasets for this purpose.

Figure 5 shows that when using a single NV-tree, recall is relatively low.
Close to 54% of the 248,212 ground truth features are found when they are lost
in the 30M collection. This percentage then slowly decreases as the distracting
collection grows, to about 38% when challenged by the 28.5B collection.

Using additional indices dramatically improves performance, however. With
the 30M collection, recall jumps to 72% with two NV-trees and 79% using three
NV-trees. At the other end of the figure, with the 28.5B collection, recall is
lower as before but remains remarkably good given the size of the distracting
collection: 52% with two NV-trees and 58% with three NV-trees.

Using more than three NV-trees provides a slight recall improvement, but not as dramatic as going from a single NV-tree to two and three. Multiplying the number of NV-trees is therefore not a worthy option, since it increases the pressure on storage and main memory and increases the retrieval cost.

4.3 Retrieval Performance

We now turn to the retrieval performance. We ran this experiment on the same system as in the previous experiment, and therefore the retrieval performance was identical. Recall that the main memory of our server was 144 GB, which means that all the leaves of three NV-trees can fit into memory for all collections except the 28.5B collection. When the various indices entirely fit in main memory, answering each query feature is extremely fast, as before: it takes a fraction of a millisecond to process one feature against one NV-tree. When using the 28.5B collection, on the other hand, main memory can not fit even one NV-tree, and the response time is therefore larger, but still only about 22.5 ms per query.

5 Experiment 3: The Small-Scale SIFT1B Benchmark

In this experiment we examine the SIFT1B benchmark, as this is the largest benchmark for which there are results for the primary competitors from the literature: Product Quantization (PQ) [7] and the Inverted Multi-Index (IMI) [2, 3]. Note, however, that we consider SIFT1B to be a relatively small benchmark. As the NV-tree index was designed primarily for very large collections, this analysis favors its competitors significantly.

Previous work has directly compared the NV-tree to many approaches, including LSH, median rank aggregation and the Spill-Tree, showing that NV-tree significantly outperforms those approaches already at a much smaller scale [8]. But it is insightful to determine the disk space requirements and expected disk read performance if the more recent state-of-the-art techniques were used to index the 28.5B features used in the previous two experiments. We make this comparison at the end of this section.

5.1 Experimental Setup

SIFT1B is a collection of exactly one billion 128-dimensional SIFT features extracted from natural images [7], which is publicly available and has been used in many publications, including [3]. The dataset comes with pre-calculated ground truth, where the exact 1,000 nearest neighbors for each of the 10,000 queries are provided; these neighbors were identified using a (long!) sequential scan computing euclidean distances. The measure used to compare systems indexing this SIFT1B dataset is the Recall@R, which for varying values of R determines the average rate of queries for which the 1-nearest neighbor is ranked in the top R positions. The SIFT1B data set was indexed by the NV-tree and the queries run. We then compare those results to the results reported in [3],

Table 3. Performance comparison using SIFT1B. Results for PQ and IMI variants are reproduced from [3].

Indexing method	Bytes/feature	Quality (Recall@1)	Features read	Retrieval time	Index size
NV-tree (3 indices)	18	0.076	3 K	1.2 ms	18 GB
PQ	12	0.112	~8 million	155 ms	12 GB
Multi-index	12	0.158	10 K	2 ms	13 GB
	12	0.165	100 K	11 ms	13 GB
- OMulti-D-OADC-local	12	0.268	10 K	6 ms	15 GB
	12	0.286	100 K	50 ms	15 GB
- 16 bytes per feature & OMulti-D-OADC-local	20	0.421	10 K	7 ms	23 GB
	20	0.467	100 K	66 ms	23 GB

which were obtained using fairly similar hardware. In our analysis, we focus on the Recall@1 measure, and observe that all collections fit easily in main memory, as we use the same server as before.

5.2 Results

Table 3 shows the comparison of the three methods: PQ, IMI, and NV-tree. For IMI, three variants are shown, where the latter apply some additional optimizations [3]. As the table shows, IMI improves on PQ in all aspects: result quality, features read, and retrieval time. Optimizations to IMI then further improve quality but at a very significant cost of retrieval time.

Table 3 also shows that while the NV-tree returns worse results, it performs better, both in terms of features read and retrieval time. By construction, the NV-tree scans contents from exactly four leafs in order to build its result, and was asked to return only $k = 1,000$ neighbors from each tree, for a total of 3,000 features scanned. As the NV-tree only stores the identifier of the feature, no re-ranking is performed after retrieval, unlike IMI and PQ, which may help explain the reduced result quality. Furthermore, unlike the other approaches, only one disk access is required per index even in the large-scale scenario of the following experiment.

This comparison shows that the NV-tree and IMI offer quite different trade-offs between retrieval performance and quality. IMI requires more work to (i) discover which cells to read, (ii) read the cells, and (iii) post-process the results. At a small scale, however, such as with this small SIFT1B collection, IMI does indeed offer a good trade-off between quality and retrieval time. As the scale of the collections grows, however, this trade-off becomes less and less viable.

5.3 Scalability Analysis

It is insightful to determine the disk space requirements and expected disk read performance if PQ and IMI were used to index the 28.5B features used in the previous experiments. PQ uses at least 32 bytes per feature, so indexing 28.5B features would require close to 1TB of memory. Product quantization must scan a substantial number of cells in order to return high quality results—typically 16 cells per query feature, with each cell requiring at least one random disk read—which amounts to a few hundred thousands points at least; clearly impossible in a disk-based setting.

IMI improves on PQ in a main-memory situation, and can obtain reasonable quality by examining only between 10 thousand and 100 thousand data points. Their most compact proposal uses only 12 bytes per feature (4 for the identifier and 8 bytes for information used for improving quality). With these settings, indexing the 28.5B collection in memory would require at least 320 GB of main memory (ignoring all overheads). In a disk-based setting, however, the key question is how many cells would be read, as each cell requires a random disk read. Unfortunately, [3] gives no information on the cell size distribution. As they use $2^{14} \times 2^{14}$ cells, however, each cell would contain on average just over one hundred data points, and 95 cells would need to be read on average to retrieve 10,000 candidates, and about 950 to retrieve 100,000 candidates. Results from [4] indicated that only about half of these cells are empty, so about 50 cells would need to be read on average to retrieve 10,000 candidates, and almost 500 cells to retrieve 100,000 candidates. Note that we have used the settings for a 1B feature collection in this analysis; it is of course not clear that those settings would yield sufficient result quality with 28.5B features. What is clear, however, is that the response time of IMI would be unacceptable at this scale in an industry setting.

In contrast, as discussed below, while we typically index each collection using three NV-trees, only a single disk read is required per NV-tree and fewer than one thousand feature identifiers are typically considered per tree when constructing the approximate answer, independent of the scale of the collection.

6   Conclusion

The NV-tree is a scalable approximate high-dimensional indexing method specifically designed for visual instance search (VIS) at large scale. In this paper, we have reported on three experiments designed to evaluate the suitability of the NV-tree for VIS applications. Two of these experiments embed VIS benchmarks from the literature within collections of up to 28.5B high-dimensional features, which is the largest single-server collection reported in the literature. The results show that indeed the NV-tree performs very well for VIS applications over large-scale collections. With the small-scale non-VIS benchmark, the NV-tree has better performance but lower result quality, but our analysis shows that query processing performance of PQ and IMI would suffer with the 28.5B collection.

The NV-tree index is a proven technology, as it is already in use at Videntifier Technologies, one of the main players in the forensics arena with technology deployed at such clients as Interpol. Their search engine targets fine-grained VIS for investigations that, e.g., aim to dismantle child abuse networks. The search engine can identify very fine-grained details in still images and videos from a collection of 150 thousand hours of video, typically scanning videos at 40x real-time speed, and about 700 h of video material can be dynamically added to the index every day.

References

1. Amsaleg, L.: A database perspective on large scale high-dimensional indexing. Habilitation à diriger des recherches, Université de Rennes 1 (2014)
2. Babenko, A., Lempitsky, V.S.: The inverted multi-index. In: Proceedings of the CVPR, Providence, RI, USA (2012)
3. Babenko, A., Lempitsky, V.S.: The inverted multi-index. IEEE Trans. Pattern Anal. Mach. Intell. **37**(6), 1247–1260 (2015)
4. Babenko, A., Lempitsky, V.S.: Efficient indexing of billion-scale datasets of deep descriptors. In: Proceedings of the CVPR, Las Vegas, NV, USA (2016)
5. Douze, M., Jégou, H., Sandhawalia, H., Amsaleg, L., Schmid, C.: Evaluation of gist descriptors for web-scale image search. In: Proceedings of the CIVR, Santorini, Greece (2009)
6. Guðmundsson, G.Þ., Amsaleg, L., Jónsson, B.Þ., Franklin, M.J.: Towards engineering a web-scale multimedia service: a case study using Spark. In: Proceedings of the MMSys, Taipei, Taiwan (2017)
7. Jégou, H., Tavenard, R., Douze, M., Amsaleg, L.: Searching in one billion vectors: re-rank with source coding. In: Proceedings of the ICASSP, Prague, Czech Republic (2011)
8. Lejsek, H., Ásmundsson, F.H., Jónsson, B.Þ., Amsaleg, L.: NV-Tree: an efficient disk-based index for approximate search in very large high-dimensional collections. IEEE Trans. Pattern Anal. Mach. Intell. **31**(5), 869–883 (2009)
9. Lejsek, H., Jónsson, B.Þ., Amsaleg, L.: NV-Tree: nearest neighbours at the billion scale. In: Proceedings of the ACM ICMR, Trento, Italy (2011)
10. Liu, T., Moore, A., Gray, A., Yang, K.: An investigation of practical approximate nearest neighbor algorithms. In: Proceedings of the NIPS, Vancouver, BC, Canada (2004)
11. Lowe, D.G.: Distinctive image features from scale-invariant keypoints. Int. J. Comput. Vis. **60**(2), 91–110 (2004)
12. Moise, D., Shestakov, D., Guðmundsson, G.Þ., Amsaleg, L.: Indexing and searching 100M images with map-reduce. In: Proceedings of the ACM ICMR, Dallas, TX, USA (2013)
13. Petitcolas, F.A.P., Steinebach, M., Raynal, F., Dittmann, J., Fontaine, C., Fates, N.: A public automated web-based evaluation service for watermarking schemes: StirMark benchmark. In: Proceedings of the Electronic Imaging, Security and Watermarking of Multimedia Contents III, San Jose, CA, USA (2001)
14. Sun, X., Wang, C., Xu, C., Zhang, L.: Indexing billions of images for sketch-based retrieval. In: Proceedings of the ACM Multimedia, Barcelona, Spain (2013)

Transactional Support for Visual Instance Search

Herwig Lejsek[1], Friðrik Heiðar Ásmundsson[1], Björn Þór Jónsson[2,3], and Laurent Amsaleg[4(✉)]

[1] Videntifier Technologies, Reykjavik, Iceland
[2] Reykjavík University, Reykjavik, Iceland
[3] IT University of Copenhagen, Copenhagen, Denmark
bjth@itu.dk
[4] CNRS–IRISA, Rennes, France
laurent.amsaleg@irisa.fr

Abstract. This article addresses the issue of dynamicity and durability for scalable indexing of very large and rapidly growing collections of local features for visual instance retrieval. By extending the NV-tree, a scalable disk-based high-dimensional index, we show how to implement the ACID properties of transactions which ensure both dynamicity and durability. We present a detailed performance evaluation of the transactional NV-tree, showing that the insertion throughput is excellent despite the effort to enforce the ACID properties.

1 Introduction

Visual instance search is the task of retrieving from a database of images the ones that contain an instance of a visual query. It is typically much more challenging than finding images that contain objects of the same category as the object in the query. If the query is an image of a shoe, visual instance search does not try to find images of shoes, which might differ from the query in shape, color or size, but tries to find images of the exact same shoe as the one in the query image.

Industry is very concerned with instance search, as various real world applications require such fine-grained image recognition capabilities. Forensics is a domain of choice, where identifying tiny similar visual elements in images is key to mapping out child abuse networks, or establishing links between various terrorism-related visual materials. Very fine-grained instance search is also involved in some copyright enforcement applications.

Instance search challenges image representations as features extracted from the images must enable fine-grained recognition despite variations in viewpoints, scale, position, illumination, etc. While holistic image representations, where each image is mapped to a single high-dimensional feature vector, are sufficient for coarse-grained similarity retrieval, local features are needed for instance retrieval.

With local features, an image is mapped to a large set of high-dimensional vectors, allowing fine-grained recognition using the multitude of small visual

© Springer Nature Switzerland AG 2018
S. Marchand-Maillet et al. (Eds.): SISAP 2018, LNCS 11223, pp. 73–86, 2018.
https://doi.org/10.1007/978-3-030-02224-2_6

matches between the query instance and the candidate images from the database. Extracting powerful local features from images has been widely studied and many strategies exist to determine (i) where local features should be extracted [21,26], and (ii) what information each local feature should encode [3,22]. All these strategies, however, result in a very large set of local features per image; an instance search system handling a few million images may need to manage a few billion local features. *Scalability* of the high-dimensional indexing techniques used in the context of instance search is therefore essential. Most state-of-the-art high-dimensional indexing solutions assume that the feature vector collection can always fit in memory. Experience from the data management community and from industry shows that this assumption is not valid, as data will eventually outgrow main memory capacity. Furthermore, with SSDs emerging as a viable middle ground between memory and hard disk drives, handling data that extends beyond main memory should be reconsidered.

Image collections grow (often rapidly) as time goes by, and it is important to ensure that the instance search engines probe up-to-date collections. Very few proposals from the literature address dynamicity, however, and most of them can only expand the indexed collection through a complete reconstruction of the index. In the real world, halting a system for re-indexing is not an option; instead, new data items must be dynamically inserted into the index while the system is running. Even the recently proposed Lambda Architecture [20], which separates handling of very recent data from older data, requires dynamic consistency of index maintenance for recent data. Furthermore, resisting failures and enforcing *durability* of the indexed data is very important. Losing the features upon failure or experiencing extended downtime for reconstruction of indices are not acceptable options. Storing the high-dimensional index on disk is thus not only necessary for scalability, but also for the dynamic integrity of the index.

Key Requirements. Based on the discussion above, we have identified the following four key requirements a high-dimensional indexing solution dedicated to enabling scalable identification of similar visual instances must meet:

R1. The index must make *efficient use of all available storage resources*, main memory, solid-state devices or hard disks.
R2. The index must offer *stable query processing performance*, so that it can be used as a component of an industry-scale processing chain.
R3. The index must support *dynamic insertion methods* so that the indexed collection can grow while concurrent retrievals are performed.
R4. The index must support the *ACID properties of transactions* (Atomicity; Consistency; Isolation; and Durability), which guarantee the integrity of index maintenance, as well as correct recovery in case of system failures.

Most state-of-the-art methods, such as product quantization, focus on compressing data into main memory. Whereas these methods often provide some guarantees on quality, they neither consider updates nor provide guarantees on query processing performance, thus failing with requirements **R1** to **R4**.

We have previously proposed the NV-tree, an approximate high-dimensional indexing method that is designed from a data management perspective and can thus deal with feature collections that outgrow main memory [14,15]. When there are more image features that can fit in RAM, the NV-tree guarantees using at most a single disk read per index to get the approximate results. By providing a disk-based query performance guarantee for the NV-tree, requirements **R1** and **R2** above are satisfied.

In [14], an insertion procedure is described and evaluated for an early version of the NV-tree. Subsequent works, which propose significant improvements to the NV-tree, also discuss dynamic maintenance of the index [15,27]. In order to make the NV-tree a fully transactional dynamic index, we propose transactional index maintenance procedures, and show that the resulting extended NV-tree supports both **R3** and **R4**.

Contributions. This article makes the following contributions to the domain of scalable high-dimensional indexing for visual instance search:

1. We show how to adapt the NV-tree to the transactional processing of insertions and deletions, guaranteeing the well-known ACID properties of transactions.
2. We evaluate insertion performance in this transactional setting and show that insertion throughput is excellent: when the index fits in memory, the index can take full advantage, but even in the disk-bound case each insertion requires only a small fraction of a disk write on average.

The technology described in this article is already in use at Videntifier Technologies, one of the main players in the forensics arena with technology deployed at such clients as Interpol. Their search engine targets fine-grained visual instance search as it is used for investigations that, for example, aim to dismantle child abuse networks. The search engine can index and identify very fine-grained details in still images and videos from a collection of 150 thousand hours of video, typically scanning videos at 40x real-time speed, and about 700 h of video material can be dynamically added to the index every day.

2   Related Work

Only *approximate* high-dimensional indexing solutions remain efficient at very large scale. Approximate indexing methods trade quality off for response time, and follow three different major directions. Due to space constraints we outline only the most scalable of these methods here; for more details see [16].

Quantization. One line of work is based on indexing data clusters such as the hierarchical k-means decomposition of the data collection: Voronoi cells are created to partition and store the high-dimensional vectors, and the cells are organized as a multi-level tree to facilitate traversal and improve response time [7].

Many variants of this basic idea have been proposed (e.g., see [17,29,30]). One algorithm from this category has been extended to cope with collections of up to 43 billion feature vectors, using distributed processing with "big data" techniques such as Spark [9,24]. A more sophisticated indexing method, still using data clusters at its core, is called product quantization [10]. Product quantization decomposes the high-dimensional space into low-dimensional subspaces that are indexed independently. This produces compact code words representing the vectors that, together with an asymmetric approximate distance function, exhibit good performance for a moderate memory footprint. Several variants of product quantization have been published; in particular, Sun et al. [32] proposed an indexing scheme based on product quantization that uses ten computers to fit in memory the 1.5 billion images collection they index. The inverted multi-index by Babenko and Lempitsky [1] uses product quantization at its core but achieves a much denser subdivision of the space by using multiple inverted indices. The experiments reported in [1], using the BIGANN dataset [11] that contains one billion SIFT descriptors, show that the approach can determine short candidate lists with superior recall. They have recently extended their method for deep learning features [2]. None of these methods support either dynamic updates or ACID properties of transactions.

Hashing. A second line of work developed around the idea of hashing. The earliest notable hashing-based method proposed was Locality Sensitive Hashing (LSH) [5]. Essentially, LSH uses a large number of hashing functions to project the high-dimensional vectors onto segmented random lines. At query time, these hash tables are probed with the query vector, and the candidates from all hash tables are then aggregated to find the true neighbors. The performance of such hashing schemes is highly dependent on the quality of the hashing functions. Hence, many approaches have been proposed to improve hashing [12,28,35]. As with the quantization-based methods, none of these hashing-based methods support either dynamic updates or ACID properties of transactions.

Tree Structures. A third approach is based on the idea of a search tree structure. The NV-tree is one proponent of this group. Fagin et al. [6] introduced the concept of median rank aggregation. They project the entire data collection on multiple random lines and index the ranked identifiers of the data points along each line, discarding the actual feature vectors. This ranking turns the high-dimensional vectors into simple sets of values which are inserted to B^+-trees. These B^+-trees are probed at search time, and the nearest neighbors of the query are returned according to their aggregated rankings. The major drawback of that algorithm is the excessive search across the individual B^+-trees [6].

Tao et al. [33] proposed another method for accessing high-dimensional data based on B^+-trees, called the locality sensitive B-tree or LSB-tree. The LSB-tree approach inherits some of the properties of LSH, but in addition projects the hashed points onto a Z-order curve. Quality guarantees can be enforced using multiple LSB-trees in combination, forming an LSB-forest [33].

Muja and Lowe [25] proposed, via the FLANN library, a series of high-dimensional indexing techniques based on randomized KD-trees, k-means indexing and random projections. Another approach in the category of search trees is the Metric tree [34]; a variant named Spill-tree is a tree-structure based on splitting dimensions in a round-robin manner, and introducing (sometimes very significant) overlap in the split dimension to improve retrieval quality [18].

Many of the tree-based methods are either based on B$^+$-trees, which support dynamic updates and transactions, or have proposed specific methods to address updates. Most of these methods, however, have only been used for relatively small collections. As far as we are aware, this article presents the first study that is considering updates at a scale of a billion features or more, and also the first study to consider transactional properties at this scale.

3   The NV-tree

The NV-tree [14, 15] is a disk-based high-dimensional index, based upon a combination of projections of data points to lines and partitioning of the projected space. By repeating the process of projecting and partitioning, data is separated into small partitions which can be easily fetched from disk with a single read, and which are likely to contain all the close neighbors in the collection. We briefly describe the NV-tree creation process, its search procedure, its dynamic insert process and then enumerate some salient properties of the NV-tree.

Index Creation. Overall, an NV-tree is a tree index consisting of a hierarchy of small *inner nodes*, which guide the vector search to the appropriate leaf node, and larger *leaf nodes*, which contain references to actual vectors. The leaf nodes are further organised into *leaf-groups* that are disk I/O units, as described below.

When tree construction starts, all vectors from the collection are first projected onto a single projection line through the high-dimensional space ([14] discusses projection line selection strategies). The projected values are then partitioned in 4 to 8 partitions based on their position on the projection line. Information about the partitions, such as the partition borders along the projection line, forms the first inner node of the tree—the root of the tree. To build the subsequent levels of the NV-tree, this process of projecting and partitioning is repeated recursively for each and every partition, using a new projection line for each partition, thus creating the hierarchy of smaller and smaller partitions represented by the inner nodes.

At the upper levels of the tree, with large partitions, the partitioning strategy assigns equal distance between partition boundaries at each level of the tree. The partitioning strategy changes when the vectors in the partition fit within 6×6 leaf nodes of 4 KB each. In this case, all the vectors from that partition are partitioned into a *leaf-group* made of (up to) 6 inner nodes, each containing (up to) 6 leaves. In this leaf-group, partitioning is done according to an equal cardinality criterion (instead of an equal distance criterion). Finally, for each leaf node, projection along a final random line gives the order of the vector identifiers and the ordered identifiers are written to disk. It is important to note that the vectors themselves are *not* stored; only their identifiers.

Indexing a collection of high-dimensional vectors with an NV-tree thus creates a tree of nodes keeping track of information about projection lines and partition boundaries. All the branches of the tree end with leaf-groups with (up to) 36 leaf nodes, which in turn store the vector identifiers.

Nearest Neighbor Retrieval. During query processing, the search first traverses the hierarchy of inner nodes of the NV-tree. At each level of the tree, the query vector is projected to the projection line associated with the current node. The search is then directed to the sub-partition with center-point closest to the projection of the query vector until the search reaches a leaf-group, which is then fully fetched into RAM, possibly causing one single disk I/O. Within that leaf-group, the two nodes with center-point closest to the projection of the query vector are identified. The best two leaves from each of these two nodes are then scanned in order to form the final set of approximate nearest neighbors, with their rank depending on their proximity to the last projection of the query vector. The details of this process can be found in [15].

Whereas the NV-tree is stored on disk, the hierarchy of inner nodes is read into memory once query processing starts, and remains fixed in memory. The larger leaf nodes, on the other hand, are read dynamically into memory as they are referenced. If the NV-tree fits into memory, the leaf nodes remain in memory and disk processing is avoided, but otherwise the buffer manager of the operating system may remove some leaf nodes from memory.

Insertions. Insertion to NV-tree leaf nodes proceeds as follows. First, the leaf node where to insert a new vector identifier is identified. The position within that leaf is also determined and the insert is performed if the leaf is not full. As for most dynamic data structures, leaf nodes at index creation time are not filled completely (they are between 50% and 85% full, and about 70% full on average) in order to leave space for such insertions. A filled leaf node must be split in order to provide more storage capacity within the tree. During a split operation, a leaf-group is considered as a unit, and all the features of the leaf-group are re-organized using the same process as during index construction. In particular, when the size of the leaf-group exceeds the capacity of 6×6 leaf nodes, the group is split into 4 to 8 new leaf-groups, depending on the distribution of the features.

During leaf-group re-organization, new projection lines are chosen for internal nodes and the new leaf nodes. As leaf nodes only contain vector identifiers, the vectors must be retrieved from disk for re-projection. In [13], it is shown that the most efficient option for handling re-projections is to maintain an independent feature database for each NV-tree, organized in the same manner as the leaf-groups, to directly read the relevant features.

Properties of NV-trees. The experiments and analysis of [15] show that the NV-tree indexing scheme has the following properties:

- *Random Projections and Ranking:* The NV-tree uses random projections to turn multi-dimensional vectors into single-dimensional values indexed by B^+-trees. Efficient implementations of dynamic B^+-trees are well known. The NV-tree does not fetch full vectors from disk to compute distances. In contrast, ranking is used, which basically amounts to scanning a list.
- *Single Read Performance Guarantee:* As leaf-groups have a fixed size, the NV-tree guarantees query processing time of a single read regardless of the size of the vector collection. Larger collections need deeper NV-trees but the intermediate nodes fit easily in memory and tree traversal cost is negligible.
- *Compact Data Structure:* The NV-tree stores in its index the identifiers of the vectors, not the vectors themselves. This amounts to about 6 bytes of storage per vector on average. The NV-tree is thus a very compact data structure. Compactness is desirable as it maximizes the chances of fitting the tree in memory, thus avoiding disk operations.
- *Consolidated Result:* Random projections produce numerous false positives that can be almost all eliminated by an ensemble approach. Aggregating the results from a few NV-trees, which are built independently over the same collection, dramatically improve result quality.

The NV-tree, however, is not a transactional index. The next section specifies mechanisms to enforce the ACID properties of transactions within the NV-tree.

4 Transactional NV-tree

Large collections of media objects, and the corresponding collections of high-dimensional vectors, are typically dynamic and require efficient insertions. For the typical web-scale application of visual instance search, however, it is safe to assume that (a) updates are made centrally, and (b) that throughput is more important for this update thread than response time. For these applications, it is feasible to batch insertions such that only one insertion thread is running at each time, which simplifies the implementation of the insertion process. Of course, however, insertions and searches must run concurrently.

This section focuses on insertions to the NV-tree index. We outline the insertion operation, then describe enforcement of the ACID properties of transactions (Atomicity; Consistency; Isolation; and Durability), and finally consider the correctness and performance of the proposed method. Note that while deletions will be rare in practice, they can be implemented using techniques very similar to those implementing insertions. We therefore briefly describe the differences for deletions, where appropriate. Updates are implemented as feature deletions followed by feature insertions.

Due to serialization of inserts, two insertion transactions will never conflict, which means that a simple locking mechanism based on tree-traversals is sufficient to enforce *isolation*. Because insertions are never aborted and they never deadlock, ensuring *atomicity* is only needed when the system crashes. Furthermore, since at most one insert transaction is running concurrently, enforcing *durability* is greatly simplified. Finally, since there are no constraints on the vectors, as such, the notion of *consistency* simply implies that the results always reflect the status after the last committed transaction.

We start by considering isolation and consistency for a single NV-tree. We then consider atomicity and durability, before addressing some practical issues relating to using multiple NV-trees.

Isolation and Consistency. Isolation is implemented by adapting a standard locking algorithm from the B$^+$-tree literature [8,31]. A search thread starts by obtaining a read lock on the root of the NV-tree. Before accessing a child node, the thread must obtain a read lock on that node. At that point, the lock on the parent can be released. Finally, the leaf-group selected for retrieval is locked and only released after all necessary identifiers have been retrieved from the leaves. Note that locks are implemented using pthread mutexes; each internal node contains the mutexes for all its children and the leaf-groups are locked as a unit since they are treated as a unit during both retrieval and node splits. As the overhead of obtaining mutexes is low, locking is always activated.

The insertion process uses the same locking mechanism, except that finally an exclusive lock is acquired for the leaf-group, preventing concurrent insertions into that leaf-group, as well as concurrent retrieval from the leaf-group. In the case of a leaf-group split, a new internal node is created pointing to all the newly created leaf-groups; the lock on the original leaf-group is sufficient to protect the modification of the parent node.

Since each query or insertion transaction needs to access multiple trees multiple times, it is necessary, however, to consider the overall interaction between search and insertion transactions. Recall that insertion transactions are serialized; they are therefore assigned with ever-increasing transaction identifiers (TIDs) that are logged with each inserted vector. Isolation is then enforced by omitting from the query result vectors with transaction identifiers larger than that of the last transaction that committed before the search started; this also guarantees consistency of the result.

Deletions are implemented in the same manner as insertions, except that a list of deleted media items is maintained to avoid returning partially deleted items; when all feature vectors from a media item have been deleted, it can be removed from this list.

Atomicity and Durability. For atomicity and durability, we adopt the standard write-ahead logging (WAL) protocol [23]. The WAL protocol uses a transaction log (or write-ahead log) which contains sufficient information to recover in case of failures. The WAL protocol has two rules to ensure the correctness of transactions:

1. The log entry for any modification must be written to disk before the modified data is written to disk.
2. All log entries for a transaction must be written to disk before the transaction can be committed.

The first rule—sometimes called the *undo* rule—ensures that any change written to the disk before it is committed can later be removed from the database, thus supporting atomicity. The second rule—the *redo* rule—ensures that committed changes can be redone in case of crashes, thus ensuring durability.

Each split can result in a large number of disk operations. Splits are therefore heavily buffered. It is best for performance to manage multiple log files where log records can be appended independently and in parallel. There is one log file per NV-tree, plus a global log file for the correctness of the overall recovery process.

The recovery manager uses regular *checkpoints* to facilitate efficient recovery. During recovery, the latest checkpoint file is first read and the status at the time of the checkpoint is adopted for the internal nodes, the leaf nodes and the leaf-group DB. Then the split operations are retrieved from the index log file, and those split operations that were performed due to committed transactions are re-played on the internal structure, while other split operations are ignored. At this point, the internal structure is correct, as of the time of the crash, but vectors may be incorrectly included and/or missing. Next, therefore, vectors that belonged to uncommitted transactions, but made it to the leaf nodes of the NV-tree are removed; note that no such vectors are ever found in the leaf-group DB, because they are only added to the leaf-group buffer when the transaction is ready to commit and the checkpoint is only written after commit. Finally, the vector collection log file is used to re-insert the committed vectors that did not make it to disk, both to the NV-tree and the leaf-group DB, taking care to avoid re-insertion to the split leaves.

Note that since insertion operations are serialized and do not conflict, the undo and redo phases can be performed in any order. Since vector removal requires moving other vector identifiers in the leaves, however, it makes sense to do that before inserting new identifiers.

Practical Issues with Multiple NV-trees. When inserting to multiple NV-trees, each tree should preferably be located on a separate hard drive (as should the log files) so that the full write-back capacity of the disks can be used for the leaf-group DB thread. In order to fully use the capacity of the disks, however, it is important to decouple the insertion process (as well as logging and checkpointing) for each NV-tree. Each NV-tree can thus be inserting from a different transaction, but they must all process the transactions in the same order. Since transactions may progress differently across different trees, more than one uncommitted transaction may have inserted vectors to some trees before a crash. Due to the ordering of transactions, however, the last NV-tree to finish a transaction decides the commit time and transactions will therefore commit in the same order, and all the techniques described above are unaffected by this change. Using decoupling, disk utilization was improved from about 40% up to 75% to 80%, without violating the previously described ACID properties.

Correctness and Recovery Performance. Since our techniques are built on standard building blocks from the database literature, which have been shown to enforce the ACID properties, a formal proof of correctness is beyond the scope of this paper. In the following, however, we give a brief outline of how such a proof would be structured.

A sufficient condition for enforcing isolation is *serializability*. Recall that we assume that insertion transactions stem from a single, serialized thread. Then the only conflicts that can arise are between this single insertion transaction and the (potentially many) retrieval transactions. As we use standard B^+-tree locking for the data structure consistency, which is known to enforce serializability, and further ensure that retrieval transactions can only see insertions from transactions that committed before the retrieval started, isolation is fully enforced. And, as discussed above, since there are no constraints on the vectors, isolation is sufficient to enforce consistency for the class of applications considered here.

By definition, the WAL protocol enforces both atomicity and durability. The fact that the log is stored in multiple files does not change this property, as long as sufficient information is stored in the log entries to redo operations in the correct order. As described in detail above, the recovery operations have been carefully ordered to ensure correctness. The proposed method therefore enforces both atomicity and durability.

Proving the correctness of the *implementation* of our method, on the other hand, is of course extremely difficult, if at all possible. The implementation has been tested very methodically, however, by pausing operations in certain places and crashing the computer; in all cases has recovery been successful. The recovery performance depends on the frequency of the checkpoints, but with reasonable checkpoint frequency the database is always fully recovered within a matter of minutes even with very large collections.

5   Performance of Index Maintenance

In this section we investigate the performance of dynamic inserts, while guaranteeing ACID properties, as described above. As the index experiences splits upon inserts, it is also important to verify that the evolution of the data structure does not impact the ability of the NV-tree to correctly identify nearest neighbors. We first discuss insertion throughput and then result quality.

This experiment was designed to show the two interesting cases that govern the performance of inserts. In the first case, when the index fits in RAM, inserts are done in memory and later asynchronously pushed to disks, resulting in excellent performance. The second case arises when the index is larger than memory. In this case, loading the affected data pages from disk may be required, which is not only slow but also interferes with writing back updated pages. We therefore expect this second case to show much worse performance.

To illustrate these two cases, we used a machine with only 32 GB of main memory. We used a small subset of a very large collection of images from Flickr to first compute 36 million SIFT vectors [19], which were indexed with three NV-trees. This is a tiny collection which can be indexed very quickly, and the resulting NV-trees together occupy slightly more than 500 MB. We then ran sequences

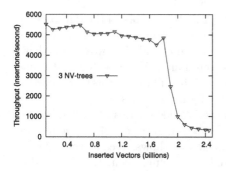

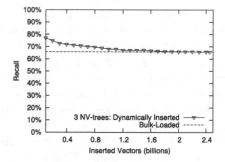

Fig. 1. Insertion throughput.   **Fig. 2.** Retrieval quality.

of 1,000 insertion transactions, where each transaction inserted 100,000 new vectors into the three NV-trees, which means that each sequence inserted 100 million new vectors. We then observed the time taken for each sequence to complete. We repeated this process and ran multiple sequences until each NV-tree contained nearly 2.5 billion vectors, occupying about 328 GB each.

Insertion Throughput. Figure 1 shows the evolution of the insertion throughput (measured by vectors inserted per second) for the duration of this workload. In the beginning of this workload, all three NV-trees fit into main memory and the throughput is excellent, around five thousand vectors per second. After running 18 such transactions, thus inserting 1.8 billion vectors, the 3 NV-trees no longer fit in main memory. After that point, Fig. 1 clearly shows the insert behavior corresponding to the second case discussed above, where the rate of inserts slows down significantly due to conflicting disk operations. It should be noted, however, that with throughput of 500 vectors per second, each insertion only takes 2 ms, which is significantly less than one disk operation per insertion, even though the descriptors are inserted to three NV-trees simultaneously.

The most important aspect of this experiment is not the reduced performance of inserts after 1.8 billion vectors have been inserted and the index no longer fits in memory; by adding more memory, larger indices can be stored in RAM. Rather it is the fact that even when the collection no longer fits in memory, and must be stored on traditional HDDs, dynamic maintenance of the index is still possible as the insertion throughput degrades gracefully. And, of course, performance of index maintenance could be vastly improved if SSDs were used instead of HDDs.

Evolution of Retrieval Quality. To evaluate the query performance of the NV-tree, we borrow the ground truth defined by [14]. A sequential scan was used to determine the 1,000 nearest neighbors of 500,000 query vectors, all coming from a very large collection of SIFTs. The resulting 500M neighbors were then analysed to identify 248,212 vectors as being meaningful nearest neighbors of the query points (as defined by [4]).

To reuse that workload, we included these 248,212 vectors in the database of 36 million other vectors used previously. Once this database was created, we ran the same 500,000 queries as in [14] and computed their recall, i.e., we counted how many of these 248,212 ground truth vectors were found. We repeated that same workload after every insertion transaction (of 100 million vectors), to observe how the quality of the answers evolves as the database grows.

Figure 2 plots the recall percentage from the 500,000 queries described above, as the collection grows in size. The figure shows a configuration where the results are aggregated from three NV-trees. As the figure shows, recall drops slowly as the collection grows, which was expected. For comparison, the figure also contains a dashed line indicating the result quality when the NV-trees for the 2.5 billion vectors are constructed from scratch via bulk loading. As the figure shows, the results for the dynamically created NV-trees and the bulk-loaded NV-trees are identical, meaning that dynamicity has no impact on result quality.

6   Conclusion

Visual instance search is the task of retrieving from a database of images the ones that contain an instance of a visual query. So far, only local image features are powerful enough to support such fine-grained recognition. Recent progress in approximate high-dimensional indexing has resulted in approaches handling several hundred million to a few billion high-dimensional vectors with excellent response times. These methods typically rely on residing in main memory for performance. We argue, however, that data quantity will always win over memory capacity in the long term. Therefore, high-dimensional indexing solutions that are truly concerned with the scalability of the feature collections they manage must address collection sizes beyond RAM capacity and efficiently utilize disks for extending storage.

Furthermore, we argue that scalability is not the only challenge that must be met as high-dimensional indexing methods must also provide dynamicity—the ability to cope with on-line insertions of features into the indexed collection, and durability—the ability to recover from crashes and avoid losing the indexed data if a failure occurs. As far as we know, no nearest neighbor algorithm published so far is able to cope with all three requirements: scale, dynamicity and durability.

In this article, we have extended an existing disk-based high-dimensional index, the NV-tree, such that it enforces the ACID properties of transactions. Experiments show that with our implementation dynamic inserts can be efficiently managed: when the index fits in memory, performance is excellent, but when the index no longer fits in memory, performance degrades very gracefully.

Indeed, the technology described in this paper is in use at Videntifier Technologies, one of the main players in the forensics arena, with technology deployed at such clients as Interpol. Their search engine is currently able to index and identify videos (at about 40x real time) from a collection of nearly 150 thousand hours of video, and about 700 h of video material can be dynamically inserted to the index every day.

References

1. Babenko, A., Lempitsky, V.S.: The inverted multi-index. IEEE Trans. Pattern Anal. Mach. Intell. **37**(6), 1247–1260 (2015)
2. Babenko, A., Lempitsky, V.S.: Efficient indexing of billion-scale datasets of deep descriptors. In: Proceedings of the CVPR, Las Vegas, NV, USA (2016)
3. Bay, H., Ess, A., Tuytelaars, T., Gool, L.V.: Speeded-up robust features (SURF). Comput. Vis. Image Underst. **110**(3), 346–359 (2008)
4. Beyer, K., Goldstein, J., Ramakrishnan, R., Shaft, U.: When is "Nearest Neighbor" meaningful? In: Beeri, C., Buneman, P. (eds.) ICDT 1999. LNCS, vol. 1540, pp. 217–235. Springer, Heidelberg (1999). https://doi.org/10.1007/3-540-49257-7_15
5. Datar, M., Indyk, P., Immorlica, N., Mirrokni, V.: Locality-Sensitive Hashing Using Stable Distributions. MIT Press, Cambridge (2006)
6. Fagin, R., Kumar, R., Sivakumar, D.: Efficient similarity search and classification via rank aggregation. In: Proceedings of the ACM SIGMOD, San Diego, CA, USA (2003)
7. Fukunaga, K., Narendra, P.M.: A branch and bound algorithms for computing k-nearest neighbors. IEEE Trans. Comput. **24**(7), 750–753 (1975)
8. Gray, J., Reuter, A.: Transaction Processing: Concepts and Techniques. Morgan Kaufmann, San Francisco (1993)
9. Guðmundsson, G.Þ., Amsaleg, L., Jónsson, B.Þ., Franklin, M.J.: Towards engineering a web-scale multimedia service: a case study using Spark. In: Proceedings of the MMSys, Taipei, Taiwan (2017)
10. Jégou, H., Douze, M., Schmid, C.: Product quantization for nearest neighbor search. IEEE Trans. Pattern Anal. Mach. Intell. **33**(1), 117–128 (2011)
11. Jégou, H., Tavenard, R., Douze, M., Amsaleg, L.: Searching in one billion vectors: re-rank with source coding. In: Proceedings of the ICASSP, Prague, Czech Republic (2011)
12. Jin, Z., et al.: Complementary projection hashing. In: Proceedings of the ACM ICCV, Barcelona, Spain (2013)
13. Jónsson, B.Þ., Amsaleg, L., Lejsek, H.: SSD technology enables dynamic maintenance of persistent high-dimensional indexes. In: Proceedings of the ACM ICMR, New York, NY, USA (2016)
14. Lejsek, H., Ásmundsson, F.H., Jónsson, B.Þ., Amsaleg, L.: NV-Tree: an efficient disk-based index for approximate search in very large high-dimensional collections. IEEE Trans. Pattern Anal. Mach. Intell. **31**(5), 869–883 (2009)
15. Lejsek, H., Jónsson, B.Þ., Amsaleg, L.: NV-Tree: nearest neighbours at the billion scale. In: Proceedings of the ACM ICMR, Trento, Italy (2011)
16. Lejsek, H., Jónsson, B.Þ., Amsaleg, L., Ásmundsson, F.H.: Dynamicity and durability in scalable visual instance search. arXiv abs/1805.10942 (2018). https://arxiv.org/abs/1805.10942
17. Li, C., Chang, E., Garcia-Molina, H., Wiederhold, G.: Clustering for approximate similarity search in high-dimensional spaces. IEEE Trans. Knowl. Data Eng. **14**(4), 792–808 (2002)
18. Liu, T., Moore, A., Gray, A., Yang, K.: An investigation of practical approximate nearest neighbor algorithms. In: Proceedings of the NIPS, Vancouver, BC, Canada (2004)
19. Lowe, D.G.: Distinctive image features from scale-invariant keypoints. Int. J. Comput. Vis. **60**(2), 91–110 (2004)

20. Marz, N., Warren, J.: Big Data: Principles and Best Practices of Scalable Real-Time Data Systems. Manning Publication co., Shelter Island (2015)
21. Mikolajczyk, K., et al.: A comparison of affine region detectors. Int. J. Comput. Vis. **65**(1), 43–72 (2005)
22. Mikolajczyk, K., Schmid, C.: A performance evaluation of local descriptors. IEEE Trans. Pattern Anal. Mach. Intell. **27**(10), 1615–1630 (2005)
23. Mohan, C., Haderle, D., Lindsay, B., Pirahesh, H., Schwarz, P.: ARIES: a transaction recovery method supporting fine-granularity locking and partial rollbacks using write-ahead logging. ACM Trans. Database Syst. **17**(1), 94–162 (1992)
24. Moise, D., Shestakov, D., Guðmundsson, G.Þ., Amsaleg, L.: Indexing and searching 100M images with map-reduce. In: Proceedings of the ACM ICMR, Dallas, TX, USA (2013)
25. Muja, M., Lowe, D.G.: Scalable nearest neighbor algorithms for high dimensional data. IEEE Trans. Pattern Anal. Mach. Intell. **36**(11), 2227–2240 (2014)
26. Nowak, E., Jurie, F., Triggs, B.: Sampling strategies for bag-of-features image classification. In: Leonardis, A., Bischof, H., Pinz, A. (eds.) ECCV 2006. LNCS, vol. 3954, pp. 490–503. Springer, Heidelberg (2006). https://doi.org/10.1007/11744085_38
27. Ólafsson, A., Jónsson, B.Þ., Amsaleg, L., Lejsek, H.: Dynamic behavior of balanced NV-trees. Multimed. Syst. **17**(2), 83–100 (2011)
28. Paulevé, L., Jégou, H., Amsaleg, L.: Locality sensitive hashing: a comparison of hash function types and querying mechanisms. Pattern Recogn. Lett. **31**(11), 1348–1358 (2010)
29. Philbin, J., Chum, O., Isard, M., Sivic, J., Zisserman, A.: Object retrieval with large vocabularies and fast spatial matching. In: Proceedings of the CVPR, Minneapolis, MN, USA (2007)
30. Philbin, J., Chum, O., Isard, M., Sivic, J., Zisserman, A.: Lost in quantization: improving particular object retrieval in large scale image databases. In: Proceedings of the CVPR, Anchorage, AK, USA (2008)
31. Srinivasan, V., Carey, M.J.: Performance of B-tree concurrency control algorithms. In: Proceedings of the ACM SIGMOD, Denver, Colorado, USA (1991)
32. Sun, X., Wang, C., Xu, C., Zhang, L.: Indexing billions of images for sketch-based retrieval. In: Proceedings of the ACM Multimedia, Barcelona, Spain (2013)
33. Tao, Y., Yi, K., Sheng, C., Kalnis, P.: Efficient and accurate nearest neighbor and closest pair search in high-dimensional space. ACM Trans. Database Syst. **35**(3), 20:1–20:46 (2010)
34. Uhlmann, J.: Satisfying general proximity/similarity queries with metric trees. Inf. Process. Lett. **40**(4), 175–179 (1991)
35. Zhang, D., Agrawal, D., Chen, G., Tung, A.: HashFile: an efficient index structure for multimedia data. In: Proceedings of the ICDE, Hannover, Germany (2011)

Interactive Product Search Based on Global and Local Visual-Semantic Features

Tomáš Skopal, Ladislav Peška[✉], and Tomáš Grošup

Faculty of Mathematics and Physics, SIRET Research Group, Charles University,
Malostranské nám. 25, 11800 Prague, Czech Republic
{skopal,peska,grosup}@ksi.mff.cuni.cz
http://www.siret.cz

Abstract. In this paper, we present a prototype web application of a product search engine of a fashion e-shop. Today, e-shop product metadata consist of text description, simple attributes (price, size, color, fabric, etc.) and visual information (product photo). Search engines used in e-shops mostly provide text and attribute/category interface for product filtering. In our model, we focus on the visual information applied in an interactive query-by-example scenario. The global visual descriptors may be often ambiguous and may not correspond well with the intended mental query of the user. Therefore, we proposed and evaluated model and GUI allowing user to guide the query process by selecting image regions (patches) of interest within the query. In the demo evaluation, we show that allowing user to specify relevant image patches led to a significant improvement of the results' relevance in the vast majority of tested queries.

1 Introduction

In e-commerce applications, the product search, browsing, and recommending models are vastly based on attribute filtering, category filtering, collaborative filtering, full-text/keyword search, etc. However, especially in fashion e-shops, it also makes sense to search for a desirable visual appearance that is provided by product photos [1].

1.1 Paper Contribution

In this work, we extend our previous model and application for product search engine [2]. One of the tasks, for which the original approach did not work well was generalization from specific query examples to broader categories, e.g., focusing on floral pattern clothes in general, while having only floral pattern shoes at hand. Although the intended visual feature can be indirectly specified by selecting additional query examples, suitable examples may be hard to find due to the bias of the original query and therefore, adding more examples may not increase

© Springer Nature Switzerland AG 2018
S. Marchand-Maillet et al. (Eds.): SISAP 2018, LNCS 11223, pp. 87–95, 2018.
https://doi.org/10.1007/978-3-030-02224-2_7

the discriminative power of the query set. Similar problem appears also in case of the items with multiple distinctive visual features, while user focuses only on a single, possibly non-dominant one.

Hence, the main contribution of the model is multi-modal approach to measuring image similarity. In addition to the original model employing global descriptor for each product image, we also extract descriptors for sub-images (patches) arranged in a regular grid.

2   Related Work

In this section we point out several research results, important for our work.

2.1   Generic CNN Models in Specific Domains

Fischer et al. [3] have shown that features obtained from deep convolutional neural networks (CNN) such as AlexNet work well when they are applied to datasets or recognition tasks different from those they were trained on. This fact is crucial in our approach, as we aim at re-using generic CNN models (like AlexNet) in various domains, in order to avoid the costly CNN training for every product niche. Long et al. [4] have also proved that learned features outperform hand-engineered features such as SIFT even in disciplines like key-point detection.

2.2   CNN in Image Patch Matching

Image patching is a sub-discipline of computer vision focusing on searching the same instance of a real-world object captured from various viewpoints, in different lighting conditions, etc. MatchNet by Han et al. [5] implements matching of image patches via Alexnet-inspired two-tower architecture. It assumes that image patches already exist and learns a metric to match them to each other. Zagoruyko and Komodakis [6] utilized a deep neural network to learn a distance function to match image patches.

2.3   Fashion Product Search

In [7], authors used a large dataset of over 800 000 clothing images with several images per product and a rich set of hierarchical content-based features. The proposed FashionNet network utilizes landmark detection, e.g., left/right sleeve, waist etc. to better reflect similarity of clothes in different poses. We aim on generally-applicable methods not restricted to fashion domain and as we mostly deal with equally processed images with isolated background, such localization is not necessary in our case. However, in case of deeper focus on fashion domain and processing "wild images", landmark detection may be utilized as a pre-processing for patches localization.

3   Similarity Model

In this work, our goal was to implement a search engine based on latent visual attributes which were not previously known to the system. Seeing the same product, two users might have different intentions of how the search should continue, with different understanding of relevance. For example, when looking at Fig. 1, one can be interested in flower-pattern items in general, whereas another user is looking for any kind of sneakers with laces. Over the last decades, many models have been proposed to address similarity search of images, including global features, local features, and hand-engineered features to target various aspects of visual information such as texture or shape. Due to the semantic gap phenomenon, majority of the low-level features are hard to translate into abstract search intentions as thought by humans, and are not optimal for the defined task.

Fig. 1.   Example product.

Recently, deep convolutional neural networks have been proved to yield state-of-the-art effectiveness in many computer vision tasks, such as classification, recognition and retrieval. Various papers, e.g. [8], show that the layered architecture targets the semantic gap and the network learns features of increased level of abstraction, starting with low level features like corners and edges in the first convolution layers, and ending with concepts defining a complete object category such as "dog" or "flower". AlexNet, the pioneer architecture of Deep Convolutional networks for high precision image classification, contains five layers of convolution, three fully connected layers and one output layer (see the architecture in [9]). Activations of neurons during a feed-forward operation can be extracted from the network, concatenated into a vector of real numbers and treated as visual-semantic descriptors. In our work, we collected activations of all neurons in fully-connected layers and the maximal activation for each filter/kernel (i.e., a max-pooled representation) in convolutional layers.

In addition to the extraction of global features, we further focused on creating a representation of local features based on image patches. The local model has the potential of increasing recall by finding objects which are not globally similar, but still have the desired visual attribute. Local representation can be also utilized to re-rank globally similar images and therefore guide the search process towards increased similarity on desired features. To create the local features, we have divided the input product images into equally sized image patches using regular 3×4 and 6×8 grids. Due to the static dataset and a fixed domain of product images, we could rely on equally captured objects without any noisy background and centered at the same position. This means that our method is applicable for any dataset satisfying these preconditions with no required supervision. It can also be applied to other $DCNNs$. After creating the image patches, descriptors of fc6, conv3, conv4 and conv5 layers were generated. The layer selection was based on our previous experiments and excluded poorly performing layers. For the purpose of the user study, only the best performing layer, fc6, was selected. However, some preliminary results show that convolutional layers also deliver a good precision as we can see in Fig. 2.

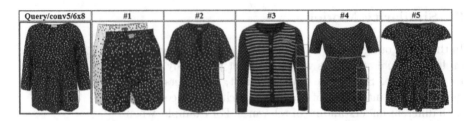

Fig. 2. Example query result obtained via search at the conv5 layer across patches created using a 6 × 8 grid.

4   Product Search Engine

The search GUI of the VADET tool[1] consists of the following elements. First, similarly as in the previous work [2], we base the search on (multiple) query-by-example paradigm allowing users to specify multiple imperfect examples that share the desired property. However, as the past experiments indicate, examples capable to sufficiently distinguish the desired features may be hard to obtain, especially if the desired features may span across heterogeneous object categories. Therefore, in addition to the selection of query examples, VADET tool allows user to specify particular image patches of interest (further denoted as local patches) for each query example (see Fig. 3f).

Following the selection of local patches, users may specify the method aggregating similarities induced by selected local patches (Fig. 3c), weight of local and global similarities in the resulting scoring function (Fig. 3b) and option to remove some patches from results, based on the percentage of background they contain (Fig. 3d).

The final scoring function s_j of object o_j is calculated as the mean of similarities $s_{i,j}$ of object o_j to each query example $o_i \in Q$. Similarity $s_{i,j}$ is defined as a weighted average of global and local similarities, where local similarity $s_{i,j}^L$ is calculated by one of following options selected by the user.

- Maximal similarity of all patch combinations:

$$s_{i,j}^L = \max_{\forall (p_{i,k}, p_{j,l}) \in S_i \times P_j} sim(p_{i,k}, p_{j,l}),$$

where S_i is the list of selected patch descriptors for object o_i, P_j is the list of all patch descriptors for object o_j, $sim(a, b)$ is cosine similarity of patch descriptors a and b. This option is particularly useful for or-like queries.

- Mean similarity of all patch combinations:

$$s_{i,j}^L = \frac{\sum_{\forall (p_{i,k}, p_{j,l}) \in S_i \times P_j} sim(p_{i,k}, p_{j,l})}{|S_i \times P_j|}.$$

This option may be useful while searching for textures filling large portions of images.

---

[1] Demo available at http://herkules.ms.mff.cuni.cz/vadet-merged.

– Mean of maximal per-patch similarity:

$$s_{i,j}^L = \frac{\sum_{\forall p_{i,k} \in S_i} \max_{\forall p_{j,l} \in P_j} sim(p_{i,k}, p_{j,l})}{|S_i|}. \tag{1}$$

This is a baseline option if the position of patches is irrelevant (i.e., searching for a specific pattern anywhere on the image).

– Mean of distance-weighted maximal per-patch similarity:

$$s_{i,j}^L = \frac{\sum_{\forall p_{i,k} \in S_i} \max_{\forall p_{j,l} \in P_j} (sim(p_{i,k}, p_{j,l}) * dist(k, l))}{|S_i|}, \tag{2}$$

where $dist(k, l)$ is an inversed normalized Euclidean distance of k and l patch centers. This option adds positional relatedness to the previous example.

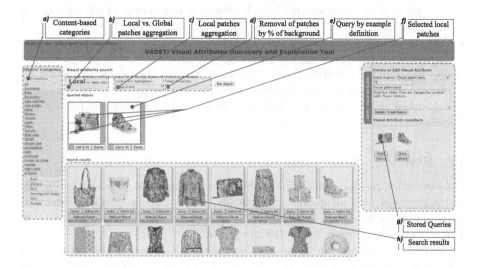

Fig. 3. Screenshot of the VADET tool with highlighted GUI options.

5   Experimental Evaluation

To measure the benefits of search based on local similarities, resp. weighting local and global similarities, a user study on our demo fashion e-shop was conducted. As this is only a fictional on-line store, we have created the basic categories and objects from the items offered by Bata and Zoot. The former is a Czech shoe and accessories manufacturer, the latter is a fashion retailer covering clothes and fashion items for all genders and age groups. Both datasets have been concatenated together (19,165 items in total) and incorporated into a category-based browsing as well as visual similarity search. Although the dataset is rather modest, it corresponds with the actual sizes of existing e-shops and therefore well reflects expected usage of the tool. We manually created 32 search tasks resembling a real shopping process with a specific intention in user's mind. Each task comprised of a description of the shopping intention, and one to three member objects to initialize the search process.

The evaluation was conducted with four male users experienced in similarity search in general. Each of the users received a training about the system properties as well as system's backend explanation. For those reasons we consider them expert users. Within the experimental evaluation, we required users to perform following actions for each manually created task/category:

1. Execute at least one query based on global similarity as a baseline.
2. Execute at least one query based on purely local similarity with manually selected image patches. The selection should have been natural to the search intention, e.g., within the category "Watches with roman digits", it would be natural to select the part of the photo that has the roman digits.
3. Execute at least one query utilizing both search models via weighting, aiming to find the optimal setting w.r.t. overall precision.
4. After each query, the system responded with a ranked list of top-20 items. Expert users have been tasked to mark relevant objects within the response and to submit the result. To ensure consistency, the system remembered all objects marked as relevant for each individual category. Final metrics were calculated using the set of all objects ever marked as relevant for a given category.

5.1   Results

The evaluation results were compared both separately w.r.t. each searched query and aggregated over all queries. Specifically, we focused on precision and discounted cumulative gain (DCG) metrics, differences between local aggregation methods and the volume of selected image patches.

Table 1 shows the averages of precision and DCG scores across all tasks as well as the gain of local and weighted models over purely global search. It can be seen that weighted model provides 35% and 29% improvement over global search w.r.t. precision and DCG at top-20 respectively.

Table 1. Average precision and average discounted cumulative gain at 5/10/20 returned result objects, and a relative improvement over global search.

Search	P@5	P@10	P@20	P@20 Rel.	DCG@5	DCG@10	DCG@20	DCG@20 Rel.
Global	54.4%	45.3%	38.4%	100.0%	1.695	2.291	3.094	100%
PureLocal	65.6%	56.3%	43.9%	114.2%	**2.055**	2.800	3.595	116.2%
Weighted	**68.1%**	**61.9%**	**51.7%**	**134.6%**	2.038	**2.931**	**3.979**	**128.6%**

Figure 4 shows relative DCG comparison of search methods. We can see that, except of five tasks, local or weighted search outperformed global search in all cases. Looking at the actual tasks, these 5 search intentions represent the entire object and cannot be locally specified. In these cases, it follows the intuition that global search is already a good mechanism, i.e., all object's patches are relevant.

Users were also given the option to manually change the local query aggregation method. In five cases, the best results were achieved by mean of max aggregation (1) indicating irrelevance of patch positioning. In all other cases, distance weighted aggregation (2) led to the best results. On average, users selected 3.25 patches/query in local search and 2.75 patches/query in weighted search. However, when restricted to the best results only, the volume of selected patches was 1.5 patches/query in both local and weighted models. This indicates that even a small amount of user interaction may lead to good results. Full results are available from goo.gl/MDjg6s.

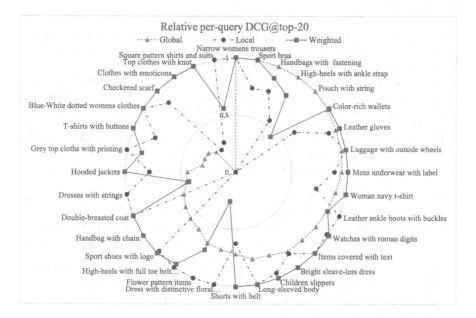

Fig. 4. Results of per-query evaluation, relative DCG w.r.t. best per-query results are depicted. Queries, for which Global method provided best results are in the top-right corner.

6   Conclusions and Future Work

Our research has shown that local interactive search based on features from existing generic neural network models can improve search engine quality significantly. Our next focus are patch extraction techniques from non-regular images, analyzing user behavior during exploration/browsing sessions and offering patch selection based on automatically discovered attributes.

We plan to utilize similarity joins [10] for better scalability and for the problem of automatic latent attributes discovery and apply evolutionary or swarm-based techniques to learn optimal patch positioning, while maintaining similar positions for items from the same category. We further plan to focus on local aggregation methods, where the proposed methods should consider aggregated deviations from the positional relations of selected patches. Another direction of our future work is to extend the proposed search model to other domains, e.g., police photo lineups assembling [11] or video retrieval [12].

Acknowledgments. This research has been supported by Czech Science Foundation (GAČR) project Nr. 17-22224S.

References

1. Hsiao, J.H., Li, L.J.: On visual similarity based interactive product recommendation for online shopping. In: 2014 IEEE International Conference on Image Processing (ICIP), pp. 3038–3041, October 2014
2. Skopal, T., Peška, L., Kovalčík, G., Grosup, T., Lokoč, J.: Product exploration based on latent visual attributes. In: Proceedings of the 2017 ACM on Conference on Information and Knowledge Management, CIKM 2017, pp. 2531–2534. ACM (2017)
3. Fischer, P., Dosovitskiy, A., Brox, T.: Descriptor matching with convolutional neural networks: a comparison to sift. CoRR, vol. abs/1405.5769 (2014)
4. Long, J., Zhang, N., Darrell, T.: Do convnets learn correspondence? CoRR, vol. abs/1411.1091 (2014)
5. Han, X., Leung, T., Jia, Y., Sukthankar, R., Berg, A.C.: MatchNet: unifying feature and metric learning for patch-based matching. In: 2015 IEEE Conference on Computer Vision and Pattern Recognition (CVPR), pp. 3279–3286, June 2015
6. Zagoruyko, S., Komodakis, N.: Learning to compare image patches via convolutional neural networks. CoRR, vol. abs/1504.03641 (2015)
7. Liu, Z., Luo, P., Qiu, S., Wang, X., Tang, X.: DeepFashion: powering robust clothes recognition and retrieval with rich annotations. In: 2016 IEEE Conference on Computer Vision and Pattern Recognition, CVPR 2016, Las Vegas, NV, USA, 27–30 June 2016, pp. 1096–1104 (2016)
8. Zeiler, M.D., Fergus, R.: Visualizing and understanding convolutional networks. In: Fleet, D., Pajdla, T., Schiele, B., Tuytelaars, T. (eds.) ECCV 2014. LNCS, vol. 8689, pp. 818–833. Springer, Cham (2014). https://doi.org/10.1007/978-3-319-10590-1_53
9. Krizhevsky, A., Sutskever, I., Hinton, G.E.: ImageNet classification with deep convolutional neural networks. In: Advances in Neural Information Processing Systems 25, pp. 1097–1105. Curran Associates Inc. (2012)
10. Čech, P., Maroušek, J., Lokoč, J., Silva, Y.N., Starks, J.: Comparing MapReduce-based k-NN similarity joins on Hadoop for high-dimensional data. In: Cong, G., Peng, W.-C., Zhang, W.E., Li, C., Sun, A. (eds.) ADMA 2017. LNCS (LNAI), vol. 10604, pp. 63–75. Springer, Cham (2017). https://doi.org/10.1007/978-3-319-69179-4_5

11. Peska, L., Trojanova, H.: Towards recommender systems for police photo lineup. In: Proceedings of the 2nd Workshop on Deep Learning for Recommender Systems, DLRS 2017, pp. 19–23. ACM (2017)
12. Lokoc, J., Bailer, W., Schoeffmann, K., Muenzer, B., Awad, G.: On influential trends in interactive video retrieval: video browser showdown 2015–2017. IEEE Trans. Multimed. (2018). https://doi.org/10.1109/TMM.2018.2830110

What Is the Role of Similarity
for Known-Item Search at Video Browser
Showdown?

Jakub Lokoč[1], Werner Bailer[2(✉)], and Klaus Schöffmann[3]

[1] Faculty of Mathematics and Physics, Charles University, Prague, Czech Republic
lokoc@ksi.mff.cuni.cz
[2] DIGITAL–Institute for ICT, JOANNEUM RESEARCH, Graz, Austria
werner.bailer@joanneum.at
[3] Institute of Information Technology, Alpen-Adria-Universität Klagenfurt,
Klagenfurt, Austria
ks@itec.aau.at

Abstract. Across many domains, machine learning approaches start to compete with human experts in tasks originally considered as very difficult for automation. However, effective retrieval of general video shots still represents an issue due to their variability, complexity and insufficiency of training sets. In addition, users can face problems trying to formulate their search intents in a given query interface. Hence, many systems still rely also on interactive human-machine cooperation to boost effectiveness of the retrieval process. In this paper, we present our experience with known-item search tasks in the Video Browser Showdown competition, where participating interactive video retrieval systems mostly rely on various similarity models. We discuss the observed difficulty of known-item search tasks, categorize employed interaction components (relying on similarity models) and inspect successful interactive known-item searches from the recent iteration of the competition. Finally, open similarity search challenges for known-item search in video are presented.

Keywords: Interactive video retrieval · Known-item search
Similarity search

1 Introduction

The age of multimedia is in full swing, with ordinary users becoming both multimedia providers and consumers. Using an ordinary smartphone, users create, share and browse more video content than ever before. In addition, cameras turn into miniature wearable devices that can capture notable moments in the life or even log the whole life [6]. In such a quantity of multimedia content, new retrieval needs and search scenarios emerge. An example is *known-item search* (KIS) in video, where users search in a given collection for a scene known a priori. Either the users saw the scene in the past (*visual KIS*) or someone described

© Springer Nature Switzerland AG 2018
S. Marchand-Maillet et al. (Eds.): SISAP 2018, LNCS 11223, pp. 96–104, 2018.
https://doi.org/10.1007/978-3-030-02224-2_8

the scene by a natural language (*textual KIS*). In both cases, the users search for one particular scene in a given video archive [9].

In order to aid with known-item search tasks in larger archives, video data have to be preprocessed, indexed for the employed similarity models [4,20], and a suitable graphical user interface has to be designed. Despite breakthroughs in machine (deep) learning [8,18], the variability and complexity of general video data represents still a challenge for automatic video retrieval approaches [1]. This is particularly the case for known-item search tasks, where users have no ideal query object at hand. Hence, trained models take just an important part in more complex decision processes interactively controlled by users. In addition, it has to be noted that user interactions are recommended also for analytical purposes (e.g., visual data mining [7]).

The user interactions during the retrieval process, however, represent a difficult challenge for evaluation and optimization of retrieval systems. User simulations can be efficiently evaluated, but their design requires observed interaction statistics and the results may deviate from the real user behavior. Evaluations with real users, on the other hand, take a lot of time and usually do not cover all aspects of the tested system and dataset. Furthermore, the presented results cannot be directly compared with other evaluations due to different conditions and settings. In order to face some of the issues, evaluation campaigns are organized [1,9,17], where teams compare their interactive retrieval systems in a competitive and shared/unified environment. In addition, the campaigns highlight promising approaches and promote the exchange of novel ideas.

In this paper, we present our experience with similarity-based approaches applied to solve known-item search tasks at the Video Browser Showdown[1] since 2015. The competition overview and observed complexity of known-item search tasks are discussed in Sect. 2. Similarity search models observed in the tools are briefly summarized in Sect. 3 and several observations of successful known-item searches from the Video Browser Showdown 2018 logs are discussed in Sect. 4. Finally, Sect. 5 presents a list of open challenges for future similarity-based known-item search systems.

2   Video Browser Showdown 2015–2018 in a Nutshell

The Video Browser Showdown (VBS) is an international evaluation campaign, where researchers from the field of multimedia come together to evaluate different approaches for interactive video retrieval on a shared dataset and a comparable basis [5,9,17]. It is performed as an extremely challenging search competition that lasts for several hours, where participating teams try to solve KIS tasks on a shared video dataset, which are evaluated in a live fashion by an evaluation server (the VBS Server). Although the main motivation for this competition is to evaluate different state-of-the-art video retrieval systems in a comparable way, the VBS is also a very entertaining event for the audience – therefore, it is organized as a side-event to the welcome reception of the MMM conference.

---

[1] www.videobrowsershowdown.org.

Table 1. Observed results of visual/textual KIS tasks at VBS 2015–2018.

Year (DB size)	2015 (100h)		2016 (250h)		2017 (600h)		2018 (600h)	
Task type	v-KIS	t-KIS	v-KIS	t-KIS	v-KIS	t-KIS	v-KIS	t-KIS
Time limit	5 min	5 min	5 min	5 min	5 min	5 min	5 min	7 min
Num. of expert tasks	10	6	10	10	7	7	4	14
Num. of novice tasks	4	2	5	0	0	0	4	0
% of solved tasks	100%	87.5%	93.3%	60%	100%	42.9%	100%	64.3%
...by the top team in category	92.9%	75%	86.7%	60%	85.7%	28.6%	75%	35.7%
% of correct submissions	63.1%	43.8%	45.7%	22.9%	54.8%	21.4%	44.4%	17.5%

Since 2015, the competition focuses on known-item search in archive collections comprising hundreds of hours of video. Since 2017, AVS tasks (find *all* relevant shots corresponding to a short textual description defined ad-hoc) are included as well.

The competition has influential effects, as the models and interfaces used by the winning tools usually affect the remaining tools for the next VBS events. Therefore, the competition aims at realistic settings of evaluated tasks, fair scoring and also includes novice user sessions to assess the usability of tools. The visual KIS tasks (i.e., searched scenes) are played to the participants on a data projector, while for the textual tasks a short text description is provided. A team solves a KIS task, if it submits a frame from the presented/observed shot (a correct submission). For more details on objectives, rules, settings and a list of participating teams/tools see [9].

Due to a fixed time slot for the event, each task has a given time limit (usually five minutes) during which all competing teams try to solve the task simultaneously. The highly competitive setting puts also more pressure on the teams and thus simulates potential stress in real situations. In connection with the difficulty of query initialization mediated just from memories, still some tasks are not solved by any of the teams and not all teams submit a correct submission for each task as presented in Table 1. For example, at the latest edition of the Video Browser Showdown in 2018 all visual KIS tasks were solved by at least one team (including novice users), while five textual KIS tasks were not solved by any of the expert users (even with an extended time limit and gradual extensions of the description). The best expert teams in the category solved only five textual KIS tasks out of fourteen. Despite limited evidence to draw significant conclusions, every year visual KIS tasks are solved by more teams than textual KIS tasks. Since 2016, the percentage of received correct submissions is approximately twice higher for visual KIS tasks. Also the average correct submission time is lower for visual KIS tasks. This corresponds to the natural expectation that observing a shot provides way more information to initialize the search and identify the correct shot in a result visualization panel than by following a textual description.

Unlike KIS tasks, in more abstractly defined AVS tasks many different correct submissions of a topic can be provided by one team and so the goal is to maximize recall provided that precision is still high. Up to one exception at VBS 2018 (one novice team in one task), all teams always found at least one correct scene for each AVS task. Again, this corresponds to a natural expectation that finding an example of a set of shots is easier than finding one particular scene in the set.

3 Similarity Models Used by Interactive Tools at VBS

Interactive multimedia search processes help in situations when the information need of the user cannot be clearly expressed. In KIS tasks this can be modeled by presenting an image or description of the target scene, which has to be translated by the user into a query (cameras are not allowed). Thus, interactive search and browsing applications make often use of a combination of steps, including similarity search approaches. Typically, low- and mid-level features can be extracted, and need to be used in the context of the content selection steps performed by the users so far. In an early work [15], a unified approach to browsing and retrieval has been proposed, with similarity-based grouping as a core component. In [2] a general framework for content abstraction is proposed, where clustering and selection are seen as the core steps in interactive browsing applications.

Based on an analysis of state-of-the-art video browsing tools [3,10,13,14], we categorize the interactions found in these tools into the following groups:

- Search by text: Searching by text in ASR or OCR transcripts, or labels/ annotations assigned automatically (e.g., by automatic concept detection).
- Search by example: Similarity search by an example image, from results or external sources, or use text as clustering feature (e.g., by topic).
- Search by sketch: Search by sketches using color, edge or motion features.
- Cluster: Cluster by similarity in terms of any of such features, or by text/labels.
- Filter: Using various attributes or actual dataset ordering, the current dataset being worked on is reduced.
- Explore: Making use of an exploration structure (e.g., similarity-based hierarchical map) to quickly inspect sets of related candidates.
- View: Inspection of video content in order to verify the relevance of a video segment. Video summarizations are often employed.

Clearly, search by an example image is the most straight-forward application of similarity search. However, also search by sketches in some feature space or clustering by these features requires a notion of similarity, as well as efficient methods for partitioning the features space around query samples or cluster centers. Finally, filtering may rely on the partitioning determined in earlier steps, or apply a threshold on the fly, again requiring identifying content segments that are within a certain distance in a feature space. In all these cases, this feature space may be multimodal, and is likely to contain features requiring different types of similarity metrics. Based on such a model, various exploration structures can be designed. Finally, text search may also include a similarity search

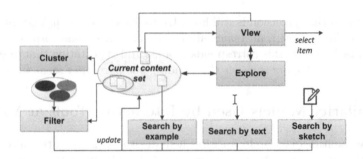

Fig. 1. Model of interactions with the content set in interactive video search.

problem, in particular when searching in noisy automatically extracted metadata. Preprocessing, such as sampling of frames from the video and performing metadata extraction, will need to be performed when ingesting content, but is not part of the interactive search process and thus not described here.

Figure 1 provides an overview how these steps can be applied iteratively in an interactive search application. The application has a current content set as its status, which can be refined by clustering, filtering or using similarity search starting from items in the set, or updated by searches using additional information (e.g., keywords, sketches).

4   Successful Known-Item Searches at VBS 2018

In the last iteration of VBS, teams were asked to provide simple unified interaction log records with each submission to the VBS evaluation server. The simplicity and instant collection have led to the first successful attempt to obtain logs from the majority of teams. As the logs were collected together with submissions, log records for many unsolved tasks are not available. Unfortunately, some teams also did not log additional utilized search interactions (e.g., browsing and video inspection). Table 2 presents a high-level overview of employed and logged query modalities for each correct submission of each team. Three types of query modalities are considered. Keyword search (K), sketch-based search (S) and query by example image (I). The letters are sorted based on their first occurrence in the log record, and the task ID is presented in the lower index[2].

The results reveal several interesting observations. Most of the successful KIS searches relied on keyword search in connection with browsing interfaces (not presented), in many cases with text query reformulation attempts. Sketch-based search often helps to solve visual KIS tasks with the clear idea of the searched scene, while example images can help with textual KIS tasks. The overall winner of the competition – namely SIRET – was the team that successfully used more

---

[2] The tasks at VBS 2018 were organized into three sessions – expert users (ID 1–12), novice users (ID 13–20) and a test session for the experts organized one day before the competition (ID 21–30).

Table 2. Solved known-item search tasks (ID in lower index) by teams at VBS 2018, including used query modalities to solve the task (K = keyword, S = sketch, I = example image, Other = filter/explore), sorted by first occurrence in the log. Teams are sorted based on the overall ranking. The HTW team did not log actions.

Team	Solved visual KIS tasks	Solved textual KIS tasks
SIRET [10]	$SKI_1, KSI_5, K_8, SKI_{11}, SIK_{15}, KSI_{17}$	$KSI_6, KI_{12}, KI_{24}, K_{25}, KI_{28}$
ITEC1 [13]	$K_8, K_{11}, K_{15}, K_{17}, K_{19}$	$K_6, K_{24}, KI_{26}, K_{30}$
ITEC2	$Other_8, KS_{13}, K_{19}$	SK_6, K_{26}
HTW [3]	Solved tasks 5, 8, 11, 19	Solved tasks 12, 22, 24, 25, 26
NECTEC [16]	K_8, K_{11}	
VIREO [12]	$S_5, SK_8, K_{11}, SK_{15}, K_{17}$	K_{22}
VITRIVR [14]	K_1, KSI_8, K_{11}	$K_{12}, K_{21}, K_{24}, K_{26}$
VERGE [11]	SI_8, K_{11}	
VNU [19]	K_5, K_{17}	K_{28}

modalities at most sessions, relying on asymmetric late fusion, simple interactive interface for query specification and video inspection [10]. This indicates the importance and help of all types of query modalities for interactive known-item video search tools. However, available logs of unsuccessful searches reveal that also other teams tried multiple modalities, i.e., the modalities are not enough.

Furthermore, the presented modalities in Table 2 have to be interpreted carefully. Not always did a modality lead directly to the result. Sometimes, it was just an investigated search option, potentially inspiring the user to reformulate the query. In Fig. 2, we can see how the expert of team ITEC1 interacted with the system solving the textual KIS task ID 26 "*Find a cartoon sailboat seen from front right, sails moving, then sails roll up and the boat is towed to an ice shell.*". First, after 18 s the user started with a textual concept search for "comic" but could not find something relevant, thus he added the search term "boat". Obviously he found something similar to the target and selected a retrieved shot for visual similarity search (even twice) but unfortunately ended up with no success. Therefore, he switched to map search for "cartoon" and "comic" (their system integrated many predefined filtered maps showing only relevant shots for a concept), and found an available map containing shots classified as "comic book". This, however, was also unsuccessful (or the user was not patient enough), therefore after browsing the map for 25 s he switched back to textual concept search for "cartoon + boat" and obviously found the right video (video 37928) after inspecting two results in more detail.

5   Open Challenges for Interactive KIS Tasks in Video

As presented in Table 1, textual known-item search tasks still represent a difficult problem even for just several hundreds hours of video. For example, the best

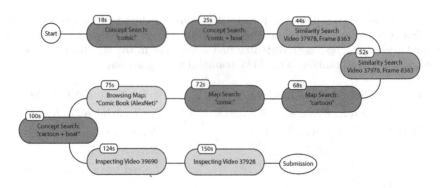

Fig. 2. The search process of team ITEC1 [13] for a KIS textual expert task.

tools at VBS 2018 were able to solve only five out of fourteen textual KIS tasks in seven minutes [9]. Despite new advances in machine learning for more effective automatic video annotation and description (affecting mostly query initialization), models for interactive similarity search will still provide complementary advanced retrieval options, Especially in cases where users have problems to express their search intents. We thus identify the following open challenges relying on similarity modeling:

- dynamic similarity models absorbing implicit/explicit relevance feedback to improve the ranking of searched scene, including indexing of dynamic models;
- similarity models for unsupervised inspection of new types of actions, taking temporal involvement and interaction of persons/objects into account;
- cross-modal similarity to help answering textual queries about visual content, by allowing the user to explore the possible space of visual representations of the elements in the query; and
- dynamic context-dependent video summarization (selection of important video frames for visualization), based on the previous browsing trail to include the aspects that are needed to judge relevance.

We believe that systems efficiently implementing the above features could push further the effectiveness of interactive known-item search systems in scenarios with limited time to solve the task.

Acknowledgments. This paper has been supported by Czech Science Foundation (GAČR) project no. 17-22224S and by grant SVV-260451. This work is also supported by Universität Klagenfurt and Lakeside Labs GmbH, Klagenfurt, Austria and funding from the European Regional Development Fund and the Carinthian Economic Promotion Fund (KWF) under grant KWF 20214 u. 3520/26336/38165.

References

1. Awad, G., et al.: TRECVID 2017: evaluating ad-hoc and instance video search, events detection, video captioning and hyperlinking. In: Proceedings of TRECVID 2017. NIST, Gaithersburg (2017)
2. Bailer, W., Thallinger, G.: A framework for multimedia content abstraction and its application to rushes exploration. In: Proceedings of the 6th ACM International Conference on Image and Video Retrieval, pp. 146–153 (2007)
3. Barthel, K.U., Hezel, N., Jung, K.: Fusing keyword search and visual exploration for untagged videos. In: Schoeffmann, K., et al. (eds.) MMM 2018. LNCS, vol. 10705, pp. 413–418. Springer, Cham (2018). https://doi.org/10.1007/978-3-319-73600-6_43
4. Chávez, E., Navarro, G., Baeza-Yates, R.A., Marroquín, J.L.: Searching in metric spaces. ACM Comput. Surv. **33**(3), 273–321 (2001)
5. Cobârzan, C., et al.: Interactive video search tools: a detailed analysis of the video browser showdown 2015. Multimed. Tools Appl. **76**(4), 5539–5571 (2017)
6. Gurrin, C., Smeaton, A.F., Doherty, A.R.: LifeLogging: personal big data. Found. Trends Inf. Retr. **8**(1), 1–125 (2014)
7. Keim, D.A.: Information visualization and visual data mining. IEEE Trans. Vis. Comput. Graph. **8**(1), 1–8 (2002)
8. Krizhevsky, A., Sutskever, I., Hinton, G.E.: ImageNet classification with deep convolutional neural networks. In: Advances in Neural Information Processing Systems, pp. 1097–1105 (2012)
9. Lokoc, J., Bailer, W., Schoeffmann, K., Muenzer, B., Awad, G.: On influential trends in interactive video retrieval: video browser showdown 2015–2017. IEEE Trans. Multimed., 1 (2018). https://doi.org/10.1109/TMM.2018.2830110
10. Lokoč, J., Kovalčík, G., Souček, T.: Revisiting SIRET video retrieval tool. In: Schoeffmann, K., et al. (eds.) MMM 2018. LNCS, vol. 10705, pp. 419–424. Springer, Cham (2018). https://doi.org/10.1007/978-3-319-73600-6_44
11. Moumtzidou, A., et al.: VERGE in VBS 2018. In: Schoeffmann, K. (ed.) MMM 2018. LNCS, vol. 10705, pp. 444–450. Springer, Cham (2018). https://doi.org/10.1007/978-3-319-73600-6_48
12. Nguyen, P.A., Lu, Y.J., Zhang, H., Ngo, C.W.: Enhanced VIREO KIS at VBS 2018. In: Schoeffmann, K., et al. (eds.) MMM 2018. LNCS, vol. 10705, pp. 407–412. Springer, Cham (2018). https://doi.org/10.1007/978-3-319-73600-6_42
13. Primus, M.J., Münzer, B., Leibetseder, A., Schoeffmann, K.: The ITEC collaborative video search system at the video browser showdown 2018. In: Schoeffmann, K., et al. (eds.) MMM 2018. LNCS, vol. 10705, pp. 438–443. Springer, Cham (2018). https://doi.org/10.1007/978-3-319-73600-6_47
14. Rossetto, L., Giangreco, I., Gasser, R., Schuldt, H.: Competitive video retrieval with vitrivr. In: Schoeffmann, K., et al. (eds.) MMM 2018. LNCS, vol. 10705, pp. 403–406. Springer, Cham (2018). https://doi.org/10.1007/978-3-319-73600-6_41
15. Rui, Y., Huang, T.: A unified framework for video browsing and retrieval. In: Image and Video Processing Handbook, pp. 705–715 (2000)
16. Rujikietgumjorn, S., Watcharapinchai, N., Marukatat, S.: Sloth search system. In: Schoeffmann, K., et al. (eds.) MMM 2018. LNCS, vol. 10705, pp. 431–437. Springer, Cham (2018). https://doi.org/10.1007/978-3-319-73600-6_46
17. Schoeffmann, K.: A user-centric media retrieval competition: the video browser showdown 2012–2014. IEEE MultiMed. **21**(4), 8–13 (2014)

18. Szegedy, C., et al.: Going deeper with convolutions. In: IEEE Conference on Computer Vision and Pattern Recognition, CVPR 2015, Boston, MA, USA, 7–12 June 2015, pp. 1–9 (2015)
19. Truong, T.D., et al.: Video search based on semantic extraction and locally regional object proposal. In: Schoeffmann, K., et al. (eds.) MMM 2018. LNCS, vol. 10705, pp. 451–456. Springer, Cham (2018). https://doi.org/10.1007/978-3-319-73600-6_49
20. Zezula, P., Amato, G., Dohnal, V., Batko, M.: Similarity Search - The Metric Space Approach. Advances in Database Systems, vol. 32. Springer, Boston (2006). https://doi.org/10.1007/0-387-29151-2

Nearest Neighbor Queries

Metric Indexing Assisted by Short-Term Memories

Humberto Razente$^{(\boxtimes)}$, Régis Michel Santos Sousa ,
and Maria Camila Nardini Barioni

Faculdade de Computação (FACOM), Universidade Federal de Uberlândia (UFU),
Uberlândia, Brazil
{humberto.razente,regismaicon,camila.barioni}@ufu.br

Abstract. Similarity queries are fundamental operations for applications that deal with complex data. This work proposes a new approach,
called *MIA (Metric Indexing Assisted by auxiliary memory with limited
capacity)*, that can be employed to create dynamic metric access methods, such as M-trees and Slim-trees, through a short-term memory. We
propose three strategies that were evaluated with various datasets and
employing different node split policies. Experimental results show that
metric access methods built with the *MIA* approach present better distribution of the elements in the index nodes when compared to the access
methods built without it. Moreover, these results show the strategies
decrease the overlap, the number of distance calculations, the number of
disk accesses and the execution time to run k-nearest neighbor queries.

1 Introduction

Nowadays we interact with several online systems that collect complex data
from which it is meaningful to search by similarity [9], such as vectorial data,
multimedia databases, bioinformatics, time series, geographic coordinates, sensor
data, etc. In this context, efficient strategies to store, to organize and to retrieve
these data is desirable.

The structures that allow indexing and fast retrieval of complex data by
similarity are known as Metric Access Methods (MAM). These MAM are
designed to reduce the number of distance calculations and the number of disk
accesses when processing similarity query operations. There have been several
research works aiming to allow efficient similarity search on large datasets, such
as [2,7,9,11,12,15–18].

Considering MAM based on ball-partitioning methods, such as M-tree [2] and
its descendants, as new instances are inserted in the trees, the index nodes of
the structures may increase and consequently their covering radii also increase.
Therefore, the probability of overlap between the regions of node coverage radii

This work has been supported by CNPq (Brazilian National Council for Supporting
Research), by CAPES (Brazilian Coordination for Improvement of Higher Level
Personnel) and by PROPP/UFU.

ⓒ Springer Nature Switzerland AG 2018
S. Marchand-Maillet et al. (Eds.): SISAP 2018, LNCS 11223, pp. 107–121, 2018.
https://doi.org/10.1007/978-3-030-02224-2_9

is increased, which in general, results in a decrease in the efficiency of similarity queries processing. In a dynamic scenario of data generation, the distribution of data can change over time [4], which can contribute to the growth of the coverage radii and the increase on the overlap on the index nodes.

Different approaches were explored in related works to overcome these issues. Dynamic reinsertions were proposed [8] for the M-tree. The main idea is to choose an instance to be reinserted in the tree when a node is about to split. Several node split strategies were proposed [2,17,18] aiming at decreasing the overlap among nodes. There are also static solutions, such as the bulk load [21] and the slim-down algorithm [18]. The former solution can be employed to recreate a more compact tree while the latter solution can be used to reorganize the leaf pages. The use of global pivots was evaluated to allow better pruning during searches [19], as well as local pivots [11]. Some of these methods have the disadvantage of increasing the overlap locally. Although they have shown to improve the performance of the methods, they are not a general solution.

This work proposes a new indexing technique based on the hypotheses that allowing to delay the insertion of data elements into a permanent MAM node can lead to the construction of more efficient MAM. This technique, called *MIA (Metric Indexing Assisted by auxiliary memory with limited capacity)*, employs a short-term memory when processing insertion operations in a MAM. The proposed technique can be applied to both metric or multidimensional structures, such as M-tree [2] or R-tree [5]. The work presented herein considered the MAM Slim-tree [18]. The *MIA* technique is applied to the MAM insertion algorithm, and there is no need to store new information in the tree nodes. Experimental results with several real datasets show that MAM built with our technique outperform MAM built without it.

Thus, the contributions of the work described in this paper can be summarized as follows:

- *MIA* technique: a new indexing algorithm that allows enhancing both the MAM construction and the similarity querying operations;
- MAM construction: MAM built with the *MIA* technique present lower overlap on the index nodes;
- MAM querying: MAM built with the *MIA* technique allows faster execution of similarity queries.

We organized this paper as follows. In Sect. 2 we describe the fundamental concepts. Section 3 details the proposed technique. Section 4 discusses the evaluation method employed including the experimental results. Section 5 presents the final considerations.

2   Fundamental Concepts

A similarity query retrieves a set of instances from a dataset ordered by their distances to a given query instance. Metric access methods are disk-based data structures created to optimize the similarity queries processing avoiding the cost of the sequential scan. Nowadays, there are several MAM described in the scientific literature. An extensive review can be found at [14]. A landmark among them is the M-tree [2], a dynamic balanced ball-partitioning hierarchy, built in a bottom-up fashion, such as the B+tree. Several works were proposed to enhance the M-tree performance, such as the Slim-tree [18], the DBM-tree [22] and the DSAT [9].

The optimization of MAM is based on the metric spaces properties [14]. A metric space is a pair $<\mathbb{S}, \delta()>$, where $\mathbb{S}$ is a data domain, and $\delta()$ is a distance function that satisfies the following axioms for any element of x, y, z in $\mathbb{S}$: $\delta(x, x) = 0$ (*identity*); $\delta(x, y) = \delta(y, x)$ (*symmetry*); $0 \leq \delta(x, y) < \infty$ (*non-negativity*); and $\delta(x, y) \leq \delta(x, z) + \delta(z, y)$ (*triangle inequality*). Among these properties, the triangle inequality is particularly essential to discard branches of the hierarchy that indeed do not contain instance answers during the MAM traversal to solve a search operation.

Briefly, ball-partitioning hierarchies are composed of index nodes and leaf nodes. An index node contains a set of pairs of the form $<pivot, radius>$, where *pivot* is a data element and *radius* is the subtree covering radius. Each one of these pairs defines a ball that covers all the data elements in the tree branch it represents. The triangle inequality is often used to determine if there is an intersection between two balls. The leaf nodes contain the data elements.

The tree is created with an empty root node. An insertion occurs in a leaf node. When the leaf is full, a split algorithm is used to create a new node and to distribute the elements between them. The upper levels are updated recursively. One such mechanism, known as bottom-up construction, guarantees that the structure is always balanced. The insertion process starts finding out a path from the root to a leaf. More than one path can be chosen if the inserted element is covered by the radii of more than one branch representative. Examples of heuristics usually employed to select a branch (ChooseSubTree algorithm) are: the smaller distance from the inserted element and the branch representatives (mindist); or minimum occupation (minoccup); or the baseline random [18].

Insertion on leaf nodes may result in overflow. In case of overflow, a split algorithm is employed to distribute the elements between the node and a new node. The work [2] proposes the use of MinMax, an algorithm that finds a pair of representatives that splits the leaf minimizing both covering radii. Its time complexity is $O(n^3)$, where n is the number of elements of a node. The work [18] proposed the use of MST that generates a minimum spanning tree and removes its longest edge, resulting in two clusters of elements that become nodes. Its time complexity is $O(n^2 \cdot log\ n)$ on the number of elements of a node. The insertion process can result in a significant overlap among nodes regarding both the leaf level and the index level. The index traversal may not be able to prune overlapped nodes. Therefore, they decrease the MAM efficiency to run similarity queries.

The MAM evaluation methodology is based on the tradeoff between the time spent to build an index and the time spent to run a query. A valuable tool that allows us to analyze the reasons why a building strategy results in better indexes is the computation of overlap measures. Considering it is not possible to compute the volume of the intersections of generic metric spaces, [18] proposed a set of measures, called fat-factors, based on counting the elements in the intersections of pairs of overlapped nodes.

The absolute fat-factor allows the comparison of indexes that store the same set of elements in the same number of nodes. This metric evaluates the quality of an index independently of a waste of disk space. The resulting measure is a value in the range $[0, 1]$, where 0 indicates no overlap and 1 indicates total overlap. Equation 1 presents the absolute fat-factor computation, where T is an index, Ic is the number of nodes read to answer a point query, H is the height, M is the number of nodes and N is the total number of elements.

$$fat_{abs}(T) = \frac{Ic - H.N}{N} \frac{1}{M - H} \qquad (1)$$

For indexes built over the same dataset that resulted in a different number of nodes (for example, due to the choice of different node split policies), the relative fat-factor takes into consideration the minimum theoretical number of nodes $(Mmin)$ and the minimum theoretical height $(Hmin)$, as presented in Eq. 2.

$$fat_{rel}(T) = \frac{Ic - Hmin.N}{N} \frac{1}{Mmin - Hmin} \qquad (2)$$

3   Metric Indexing Assisted by a Short-Term Memory

This paper presents a new technique called *MIA* through which an auxiliary memory of limited storage capacity is used for temporary data processing when building a MAM. It can be employed in the construction of dynamic MAM, such as M-tree, and its descendants. In this work we chose the MAM Slim-tree as the baseline. The Slim-tree [18] is a MAM where each tree node is stored in a fixed-size disk page with bottom-up incremental construction. Let $S \subset \mathbb{S}$ be a dataset and $\delta()$ be a distance function. The Slim-tree leaf node is composed of an array in the form:

$$\text{leaf node} \leftarrow \text{array of } <id,\ S_i,\ \delta(S_i, S_{rep})>$$

where each tuple contains an identifier (id), the data element S_i and the distance to the node representative S_{rep} (stored in the upper level). The Slim-tree index node is composed of an array in the form:

$$\text{index node} \leftarrow \text{array of } <S_{rep_i},\ page\ id,\ radius,\ \delta(S_{rep_i}, S_{rep})>$$

where each tuple contains the element S_{rep_i} (subtree representative), the page address to the pointed node $(page\ id)$, the covering radius $(radius)$, and the distance to the upper representative $\delta(S_{rep_i}, S_{rep})$ (stored in the upper level, if exists).

The *MIA* technique was developed based on the assumption that by delaying the insertion of an element into a permanent node (disk page), it may be grouped with other more similar elements that will be inserted, thus contributing to minimize the node covering radius. Therefore, it performs the insert operation in the following manner. If the insertion in the leaf node increases its covering radius, the new algorithm inserts the element into the short-term memory, delaying the persistence of the element in the tree, which would cause an immediate overlap increase.

The short-term memory is a main memory-based data structure where its size is user predefined. To enable fault tolerance a redo log file is employed to persist the inserted data elements temporarily. It is important to note that future processing will be performed to insert the elements in the tree hierarchy.

The insert algorithm works as follows. Starting from the root node, if there is a node (subtree) that covers the new element, it selects the node. Otherwise, the algorithm chooses the node whose distance from the representative is smaller. If more than one node covers the new element, the algorithm employs the ChooseSubtree heuristics. This recursive search is done until it reaches a leaf node. Figure 1 presents the complete process.

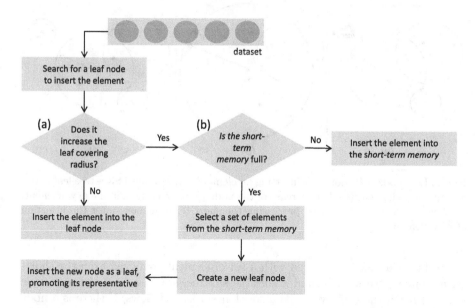

Fig. 1. Overview of the insertion algorithm assisted by the short-term memory.

After choosing the leaf node to insert the element, if the radius of the node does not need to increase and if there is space available, the algorithm adds the element. If the insertion causes overflow, it runs a split algorithm (such as the heuristics random, MinMax or MST), and the new representatives may be updated in the upper levels, recursively (Fig. 1(a)).

On the other hand, an inserted element that would lead to a leaf node radius increase is sent to the short-term memory until it reaches its capacity (Fig. 1(b)).

When there is an overflow in the short-term memory, it creates a new leaf node or a set of leaf nodes with clusters of elements removed from the short-term memory. The algorithm creates groups of elements, where each group is limited to the size of a leaf node and the occupation rate parameter. The occupation rate defines the space left in the leaf node to allow future insertions aiming to avoid a split or the creation of a new leaf node overlapped with the current node. The strategy enables a new element that is not covered by a particular leaf node to wait in the short-term memory to be grouped with other neighboring elements, thus forming a better new leaf.

The insert operation of the standard Slim-tree may lead to a leaf node increase, as shown in Figs. 2(a) and (b). In this case, the query performance of the structure decreases as the overlap increases. The *MIA* algorithm, shown in Fig. 2(c), allows the elements that would have caused the increase in the radius to form a new node. Therefore, the technique enables creating compact nodes at low processing and memory costs.

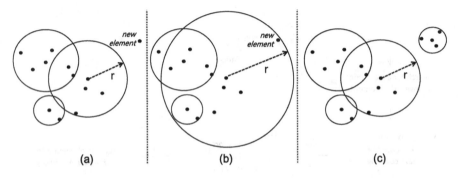

Fig. 2. Leaf node behavior after inserting an element. (a) Standard Slim-tree leaf nodes and the new element. (b) The increase of the node radius r to include the new element. (c) New index node created by processing the short-term memory after the insertion of new elements.

To insert a new leaf created with a set of elements removed from the short-term memory, we need to find a suitable subtree, starting from the root node until we reach the leaf level. The algorithm works as follows. If there is a node (subtree) that covers the new leaf, the algorithm selects it. Otherwise, if there is not a node that covers it, choose the node whose distance to the new leaf is smaller. If more than one node covers the new leaf, we propose a ChooseSubtree method based on Mindist that receives a leaf node instead of a single element. It chooses the subtree whose radius increase will be smaller. After inserting the new leaf, the insert algorithm promotes the leaf representative element. As in the standard Slim-tree, if the promotion causes overflow on the index node, it runs the split algorithm, promoting the new representatives, recursively.

We defined three strategies to get clusters of elements from the short-term memory. It is important to note that the requirement of a low processing computational cost drove the development. The goal is to select a representative element (or a set of representative elements) from the short-term memory which minimizes its sum of distances to the closest elements (limited to the node capacity and the occupation rate). The challenge is to define how to create a leaf node, considering the tradeoff between the computational cost and the ability to create compact clusters.

3.1 SM-Random

The SM-Random algorithm randomly selects a short-term memory element as the node representative. The distances from the chosen element to all the elements from the short-term memory are computed and ranked. Algorithm 1 selects the most similar elements up to the node capacity and the occupation rate.

Algorithm 1. SHORT-TERM MEMORY RANDOM LEAF NODE.

Input: The elements x in the short-term memory S ($x \in S$), the node
    capacity c, the occupation rate t
Output: Leaf node $node$

1 Randomly choose an object $x_i \in S$ to be rep
2 **for** $x_i \in S$ **do**
3 | $d[i] \leftarrow \delta(rep, x_i)$
4 **end**
5 Sort vectors d and S according to d in descending order
6 Remove the objects from S in backward order until it fills the leaf page
    according to c and t, $node \leftarrow remove(S, c, t)$
7 Return $node$

3.2 SM-Density

The SM-Density algorithm aims at finding a set of elements up to the node capacity and the occupation rate that minimizes the sum of distances to the representative element. To avoid a time complexity of $O(n^2)$ on the size n of the short-term memory, we set a constant amount of random selection of representatives, whose distances to all the elements from the short-term memory must be computed and ranked. Finally, the densest set found is selected to create the new leaf node.

3.3 SM-Cluster

The SM-Cluster algorithm employs the algorithm *CLARANS (Clustering Large Applications based on Randomized Search)* [10]. It could use other k-medoids clustering algorithms. The algorithm sets each medoid found as a leaf node representative and the closest elements from each representative, up to the node capacity and the occupation rate, are grouped. To define the number of clusters, we employed the ratio between the size of the short-term memory and the leaf node capacity, limited to a constant value, as the algorithm time complexity grows exponentially on the number of clusters.

4   Experiments

The access methods were implemented in C++ [1]. We run the experiments on a Linux 64 bits personal computer with 8 GB of main memory, Intel Core i7-4770@3.40 GHz, and 1 TB hard disk.

The following configuration parameters were active in the experiments: occupation rate of the newly created nodes from the SM: 90%; distance function: Euclidean (L_2); short-term memory: 500 elements; node split policy: Min-Max or MST algorithms; k-nearest neighbor queries: random selection of 100 query elements; SM-Density strategy: randomly selects 5% (25 elements) of the short-term memory as representatives; SM-Cluster strategy: CLARANS parameters $k = max(n/c, 5)$, $maxNeigbor = max(50, 0.0125 \cdot (k \cdot (n - k)))$ and $numLocal = 2$, where k is the number of clusters, c is the node capacity (number of elements), and n is the size of the short-term memory.

Table 1 shows the datasets evaluated in the experiments. It is important to notice the fourth column of the table shows the size allocated in KB for the short-term memory (where the size is related to the number of dimensions).

Table 1. Datasets.

Dataset	Number of elements	Number of dimensions	Short-term memory size in KBytes	Reference
Pendigits	10,992	16	32	[6]
Eigenfaces	11,900	16	32	[6]
Letter	20,000	16	32	[6]
SISAP Nasa	40,150	20	40	[3]
Corel	68,040	32	64	[6]
Covertype	581,102	54	108	[6]
SISAP Colors	112,682	112	226	[3]

The use of different node split algorithms may lead to different construction and retrieval efficiencies. In general, MinMax results in smaller overlap among nodes when compared to MST, but on the other hand, a higher number of distance calculations is necessary during insertions. Thus, the experiments evaluate the use of the new algorithms presented in this paper (SM-Random, SM-Density, and SM-Cluster) with different node split algorithms to build efficient metric access methods. We compared the SM-Random (SMR), SM-Density (SMD) and SM-Cluster (SMC) to the standard Slim-tree (Slim) [18]. For the k-nearest neighbors' experiments, we also run a sequential scan (Seq) as a baseline. In all the following experiments we highlighted in bold the smaller values (intragroup comparison).

The experiments resulted in the following information concerning the node split algorithms: the time spent to build the indexes; the indexes quality comparison; the performance of the indexes to run queries. The experimental analysis considers an intragroup comparison among the indexes built with the same split algorithm and an intergroup comparison among indexes built with different split algorithms. Three questions guided the analysis:

- $Q1$: is it possible to create indexes in smaller processing time by using a short-term memory?
- $Q2$: is it possible to build indexes with lower overlap among nodes by using a short-term memory?
- $Q3$: is it possible to run similarity queries faster on indexes built with the aid of a short-term memory?

4.1   Evaluation of the Index Construction

To answer $Q1$, "is it possible to create indexes in smaller processing time by using a short-term memory?", Table 2 shows the time in seconds to build the indexes. If we consider the intragroup comparison, the use of a short-term memory increased the time spent on indexing the data when we employed MST split algorithm for the first five datasets and slightly decreased the time spent for the two most representative datasets. An impressive result is that the use of the short-term memory allowed the indexes based on MinMax to be built faster. The intuition is that as the insertions are postponed and as the new nodes are tighter, further splits were avoided, resulting in better runtime.

Table 2. Time in seconds to build the indexes.

Dataset	MinMax				MST			
	Slim	SMR	SMD	SMC	Slim	SMR	SMD	SMC
Pendigits	0.890	**0.608**	2.087	4.275	**0.124**	0.187	1.787	3.634
Eigenfaces	0.972	**0.707**	1.875	3.808	**0.134**	0.175	1.317	3.819
Letter	1.533	**1.275**	3.968	6.433	**0.234**	0.329	3.915	6.193
SISAP Nasa	2.833	**1.978**	10.463	27.722	**0.532**	0.778	9.648	28.238
Corel	10.669	**8.689**	14.977	42.681	**1.210**	1.354	9.380	43.220
Covertype	235.854	**125.060**	211.053	632.575	19.307	**18.792**	111.047	595.054
SISAP Colors	77.684	**68.441**	74.384	142.996	5.983	**5.690**	16.951	156.441

4.2 Analysis of the Tree Overlap

To answer $Q2$, "is it possible to build indexes with lower overlap among nodes by using a short-term memory?", we computed the fat-factors as presented in Tables 3 and in 4. We also show the resulting number of nodes in Table 5. From Tables 3 and 4 it is possible to notice the proposed algorithms allows the quality of the indexes to increase. For instance, while Slim-Tree presents a relative fat-factor of 0.138 for the Covertype dataset using MinMax, SMR and SMD resulted in a relative fat-factor of 0.100 and 0.101.

Table 3. Relative fat-factor.

Dataset	MinMax				MST			
	Slim	SMR	SMD	SMC	Slim	SMR	SMD	SMC
Pendigits	0.347	0.212	**0.209**	0.220	0.381	**0.207**	0.223	0.209
Eigenfaces	0.327	**0.138**	0.166	0.152	0.359	**0.158**	0.175	0.165
Letter	0.581	**0.292**	0.296	0.297	0.633	0.344	**0.332**	0.352
SISAP Nasa	0.397	**0.212**	0.228	0.227	0.472	**0.223**	**0.223**	0.231
Corel	0.541	0.254	**0.218**	0.242	0.503	**0.239**	0.257	0.243
Covertype	0.138	**0.100**	0.101	0.108	0.129	**0.099**	0.100	0.101
SISAP Colors	0.489	0.371	**0.339**	0.346	0.438	**0.263**	0.342	0.295

To guide the further analysis, [20] empirically defined the following classification regarding the quality of an index: an index with absolute fat-factor from 0 to 0.1 is considered good, from 0.1 to 0.3 is acceptable and an index with absolute fat-factor greater than 0.3 results in performance degradation. Analysing the absolute fat-factor values obtained (Table 4) it is possible to verify the indexes built with the use of the MinMax and MST split algorithms, with the aid of the algorithms presented herein, are acceptable. Moreover, it is also possible to conclude there was a significant quality improvement on the indexes by postponing the insertions with the aid of the short-term memory. For instance, the absolute fat-factor of the indexes built for the Covertype dataset (Table 4) decreased 29.4% and 30.4% for the MinMax split and 15.2% and 13.9% for MST split, considering SMR and SMD respectively.

Table 5 shows the resulting number of nodes of the indexes. For instance, analyzing the results for the Covertype dataset, the number of nodes is 3.2% and 5.3% greater for the SMR and the SMD respectively when compared with the Slim-tree with MinMax. Although the use of the algorithms SMR, SMD and SMC resulted in trees with more nodes, the nodes are tighter, and the elements are better distributed, considering they resulted in trees with smaller absolute and relative fat-factors. On the other hand, analyzing the results obtained for the trees built with MST, it is possible to see there is a reduction of 9.9% and 10.6% on the number of nodes for SMR and SMD. In this case, the short-term

Table 4. Absolute fat-factor.

Dataset	MinMax				MST			
	Slim	SMR	SMD	SMC	Slim	SMR	SMD	SMC
Pendigits	0.234	0.137	**0.129**	0.136	0.208	**0.122**	0.127	**0.122**
Eigenfaces	0.216	0.087	0.102	**0.093**	0.189	**0.087**	0.094	0.092
Letter	0.418	0.186	**0.182**	0.186	0.332	0.193	**0.185**	0.197
SISAP Nasa	0.269	**0.134**	0.143	0.137	0.236	**0.132**	**0.132**	0.134
Corel	0.430	0.174	**0.149**	0.166	0.293	**0.149**	0.158	0.155
Covertype	0.102	0.072	**0.071**	0.078	0.079	**0.067**	0.068	**0.067**
SISAP Colors	0.422	0.300	**0.276**	0.278	0.292	**0.173**	0.233	0.202

memory allowed better distribution of the elements among the nodes, reducing the number of nodes and also decreased the overlap.

Table 5. Number of nodes in the resulting indexes.

Dataset	MinMax				MST			
	Slim	SMR	SMD	SMC	Slim	SMR	SMD	SMC
Pendigits	**581**	634	662	657	718	**692**	718	695
Eigenfaces	**642**	690	713	714	806	791	821	**781**
Letter	**990**	1141	1182	1155	1359	1298	1304	**1296**
SISAP Nasa	**2574**	2791	2815	2907	3484	**2963**	2979	3027
Corel	**2951**	3443	3453	3434	4018	3775	3827	**3700**
Covertype	**21796**	22503	22960	22425	26300	23688	**23501**	24557
SISAP Colors	**3732**	4009	3968	4025	4833	4910	4757	**4731**

Considering the SISAP Nasa dataset, the results presented in Table 3 show the increase in the quality of the indexes. When compared to the standard Slim-tree using MinMax we have an increase in the quality of the indexes of 46.6% for SMR and 42.6% for SMD and the results shown for MST we have an increase in the quality of the indexes of 52.8% for both SMR and SMD. The absolute fat-factor results shown in Table 4 also allows verifying that the indexes built with MinMax and MST split algorithms considering the MIA technique are acceptable. Moreover, the experimental results presented in Table 5 shows the proposed algorithms allows to reduce the number of nodes of the indexes built with the MST split algorithm.

4.3 Performance of Similarity Queries

To answer $Q3$, "is it possible to run similarity queries faster on indexes built with the aid of a short-term memory?", we run 100 k-nearest neighbor queries with $k = 100$. The query elements were selected randomly. It is important to notice that to run the queries the short-term memory may have data elements, the k-nearest neighbor algorithm described in [13] was adapted to search first the short-term memory.

Table 6 presents the query processing times. An impressive result is that the SMR, SMD and SMC indexes built with both MinMax and MST achieve very similar query processing times, making the MST feasible, as MST decreases the time spent to build an index when compared to MinMax. Thus, MST can now be used to enhance the tradeoff between construction and query times, reducing the build time and achieving query processing times comparable to that of the access methods built with MinMax.

Table 6. Total time in seconds to run 100 executions of 100-nearest neighbors' queries.

Dataset	MinMax					MST				
	Seq	Slim	SMR	SMD	SMC	Seq	Slim	SMR	SMD	SMC
Pendigits	0.177	0.090	0.054	**0.052**	**0.052**	0.177	0.088	0.053	0.052	**0.051**
Eigenfaces	0.188	0.072	**0.047**	0.051	0.048	0.188	0.078	0.049	0.053	**0.048**
Letter	0.309	0.262	**0.156**	0.161	0.159	0.309	0.264	0.161	**0.160**	0.162
SISAP Nasa	0.751	0.454	**0.261**	0.280	0.280	0.751	0.468	**0.251**	0.267	0.271
Corel	1.538	0.951	0.571	**0.517**	0.552	1.538	0.868	**0.515**	0.546	0.527
Covertype	18.661	2.417	1.278	**1.079**	1.229	18.661	1.895	1.134	**0.956**	1.040
SISAP Colors	7.473	4.144	3.314	**3.192**	3.265	7.473	3.202	**2.810**	3.065	2.892

Table 7 shows the average number of disk accesses and Table 8 shows the average number of distance calculations. These experimental results are somehow related to the query processing time as the increase in one or both measures directly increases the query processing time. They are also subject to the number of dimensions of each dataset, as the higher the value, the higher the time to process each distance and the higher the disk space used. The curse of dimensionality [14] corroborates to the difficulty of searching or mining high dimensional data as the distances between pairs of elements tend to be very similar, decreasing the performance of the algorithms.

Analysing the total number of nodes of the MinMax indexes shown in Table 5, although SMR, SMD, and SMC resulted in indexes with more nodes than the standard Slim-tree built with MinMax, the number of nodes visited during k-nearest neighbor queries is smaller (Table 7). Considering MST indexes, although SMR, SMD, and SMC resulted in indexes with more nodes than the standard Slim-tree built with MinMax and also SMR, SMD, and SMC built with MinMax, the number of nodes visited during k-nearest neighbor queries is comparable to the same experiments with MinMax. These results emphasize that the *MIA* technique allowed the indexes built with MST to achieve the query processing times similar to that of the indexes built with MinMax.

Table 7. Average number of disk accesses of 100-nearest neighbors' queries.

Dataset	MinMax					MST				
	Seq	Slim	SMR	SMD	SMC	Seq	Slim	SMR	SMD	SMC
Pendigits	423	344	234	**220**	227	423	360	235	232	**226**
Eigenfaces	458	263	**189**	211	200	458	300	**207**	228	**207**
Letter	770	871	**758**	776	787	770	992	824	**812**	833
SISAP Nasa	1825	1663	**1327**	1388	1413	1825	1933	**1320**	1359	1399
Corel	2347	2177	1695	**1528**	1633	2347	2178	**1614**	1682	1617
Covertype	16601	4999	3007	**2607**	3067	16601	3889	2741	**2446**	2723
SISAP Colors	3220	2593	2543	**2432**	2502	3220	2659	**2433**	2573	**2433**

By analyzing the results presented in Table 8, it is possible to notice the reduction of 51.5% from standard Slim-tree to SMD (MinMax) for the Covertype dataset (intragroup comparison). Considering the intergroup comparison it is possible to observe that the queries run on MST indexes for SMR, SMD and SMC resulted in fewer distance calculations than the same methods built with MinMax for all datasets.

Table 8. Average number of distance calculations of 100-nearest neighbors' queries.

Dataset	MinMax					MST				
	Seq	Slim	SMR	SMD	SMC	Seq	Slim	SMR	SMD	SMC
Pendigits	10992	4496	3282	**3119**	3152	10992	4213	3197	3113	**3070**
Eigenfaces	11900	3456	2902	3138	**2885**	11900	3423	2903	3163	**2873**
Letter	20000	14976	**11087**	11347	11474	20000	14170	**11062**	11150	11301
SISAP Nasa	40150	21433	**14650**	15897	15730	40150	21025	**13964**	15344	15489
Corel	68040	37425	25779	**23181**	24750	68040	32549	**22619**	24294	23444
Covertype	581012	59108	34444	**28687**	31643	581012	46415	29982	**24536**	26289
SISAP Colors	112682	60130	56619	**54789**	56237	112682	54221	**47017**	51461	48575

5   Conclusion

The development of efficient dynamic metric access methods is fundamental for similarity search. We developed a technique called *MIA* that employs a limited amount of memory to improve the performance of these methods. *MIA* is generic and may be used by several ball-partitioning metric access methods and by multidimensional methods. We empirically show that it decreases the growth in the volume of the nodes and the overlap among nodes. Consequently, it speeds up similarity queries over complex data. The experimental results show the use of the short-term memory allowed to build indexes based on MinMax (split) faster.

The results also show the indexes created with MST (split) achieved the query processing times comparable to that of the indexes built with MinMax. Future work includes evaluating the method on indexes that employ cut-regions [7] instead of ball-regions.

References

1. Arboretum library. https://bitbucket.org/gbdi/arboretum. Accessed July 2018
2. Ciaccia, P., Patella, M., Zezula, P.: M-tree: an efficient access method for similarity search in metric spaces. In: International Conference on Very Large Data Bases (VLDB), Greece, Athens, pp. 426–435 (1997)
3. Figueroa, K., Navarro, G., Chávez, E.: Metric spaces library (2007). http://www.sisap.org/Metric_Space_Library.html
4. Gama, J.: A survey on learning from data streams: current and future trends. Progr. Artif. Intell. **1**(1), 45–55 (2012)
5. Guttman, A.: R-trees: a dynamic index structure for spatial searching. In: International Conference on Management of Data (SIGMOD), Boston, MA, pp. 47–57 (1984)
6. Lichman, M.: UCI Machine Learning Repository. School of Information and Computer Sciences, University of California, Irvine (2013). http://archive.ics.uci.edu/ml
7. Lokoc, J., Mosko, J., Cech, P., Skopal, T.: On indexing metric spaces using cut-regions. Inf. Syst. **43**, 1–19 (2014)
8. Lokoc, J., Skopal, T.: On reinsertions in m-tree. In: International Workshop on Similarity Search and Applications (SISAP), pp. 121–128. IEEE (2008)
9. Navarro, G., Reyes, N.: New dynamic metric indices for secondary memory. Inf. Syst. **59**, 48–78 (2016)
10. Ng, R.T., Han, J.: CLARANS: a method for clustering objects for spatial data mining. IEEE Trans. Knowl. Data Eng. **14**(5), 1003–1016 (2002)
11. Oliveira, P.H., Traina, C., Kaster, D.S.: Improving the pruning ability of dynamic metric access methods with local additional pivots and anticipation of information. In: Tadeusz, M., Valduriez, P., Bellatreche, L. (eds.) ADBIS 2015. LNCS, vol. 9282, pp. 18–31. Springer, Cham (2015). https://doi.org/10.1007/978-3-319-23135-8_2
12. Razente, H.L., Lima, R.L.B., Barioni, M.C.N.: Similarity search through one-dimensional embeddings. In: ACM Symposium on Applied Computing (SAC), Marrakech, Morocco, pp. 874–879 (2017)
13. Roussopoulos, N., Kelley, S., Vincent, F.: Nearest neighbor queries. In: International Conference on Management of Data (SIGMOD), San Jose, pp. 71–79 (1995)
14. Samet, H.: Foundations of Multidimensional and Metric Data Structures. Morgan Kaufmann, San Francisco (2006)
15. Silva, Y.N., Aref, W.G., Larson, P.-A., Pearson, S., Ali, M.H.: Similarity queries: their conceptual evaluation, transformations, and processing. VLDB J. **22**(3), 395–420 (2013)
16. Skopal, T.: On fast non-metric similarity search by metric access methods. In: Ioannidis, Y., et al. (eds.) EDBT 2006. LNCS, vol. 3896, pp. 718–736. Springer, Heidelberg (2006). https://doi.org/10.1007/11687238_43
17. Souza, J., Razente, H., Barioni, M.C.: Optimizing metric access methods for querying and mining complex data types. J. Braz. Comput. Soc. (JBCS) **20**(17), 14 (2014)

18. Traina, C., Traina, A., Faloutsos, C., Seeger, B.: Fast indexing and visualization of metric data sets using slim-trees. IEEE Trans. Knowl. Data Eng. (TKDE) 14(2), 244–260 (2002)
19. Traina, C., Traina, A., Filho, R.S., Faloutsos, C.: How to improve the pruning ability of dynamic metric access methods. In: International Conference on Information and Knowledge Management (CIKM), McLean, pp. 219–226 (2002)
20. Traina, C., Traina, A., Seeger, B., Faloutsos, C.: Slim-trees: high performance metric trees minimizing overlap between nodes. In: International Conference on Extending Database Technology (EDBT), Konstanz, pp. 51–65 (2000)
21. Vespa, T., Traina, C., Traina, A.: Efficient bulk-loading on dynamic metric access methods. Inf. Syst. 35(5), 557–569 (2010)
22. Vieira, M.R., Traina, C., Chino, F.J.T., Traina, A.: DBM-tree: a dynamic metric access method sensitive to local density data. J. Inf. Data Manag. (JIDM) 1(1), 111–127 (2010)

New Permutation Dissimilarity Measures for Proximity Searching

Karina Figueroa[1]([✉])[iD], Rodrigo Paredes[2][iD], and Nora Reyes[3][iD]

[1] Universidad Michoacana, Morelia, México
karina@fismat.umich.mx
[2] Universidad de Talca, Curicó, Chile
raparede@utalca.cl
[3] Universidad Nacional de San Luis, San Luis, Argentina
nreyes@unsl.edu.ar

Abstract. Proximity searching consists in retrieving the most similar objects to a given query from a database. To do so, the usual approach consists in using an index in order to improve the response time of online queries. Recently, the *permutation based algorithms (PBA)* were presented, and from then on, this technique has been very successful. In its core, the PBA uses a metric between permutations, typically Spearman Footrule or Spearman Rho. Until now, several proposals based on the PBA have been developed and all of them uses one of those metrics. In this paper, we present a new family of dissimilarity measures between permutations. According to our experimental evaluation, we can reduce up to 30% the original technique costs, while preserving its exceptional answer quality. Since our dissimilarity measures can be applied in any state-of-the-art PBA variant, the impact of our proposal is significant for the similarity search community.

Keywords: Approximate similarity searching
Permutation based algorithms · Permutation dissimilarity measures

1 Introduction

Proximity searching is the core of many applications in numerous fields, such as, image classification, audio, document databases, compression, computational biology, information retrieval, and data mining, to mention a few. All these applications have in common a database, usually huge, and a metric distance function between any two elements, generally expensive to compute. The basic approach is to build an index over the database in preprocessing time, so that subsequent queries can be answered efficiently. The goal is to avoid comparing the query against the whole database. Basically, there are two fundamental types of queries: (i) $k-nearest\ neighbors$ or k–NN, that retrieves the k most similar elements to the query, and (ii) *range query*, that retrieves the set of elements within a fixed distance r to the query.

© Springer Nature Switzerland AG 2018
S. Marchand-Maillet et al. (Eds.): SISAP 2018, LNCS 11223, pp. 122–133, 2018.
https://doi.org/10.1007/978-3-030-02224-2_10

There are many effective algorithms in low dimensional spaces, however, their performance worsen as the intrinsic dimension increases, which is known as the *curse of dimensionality* [5]. In fact, in high dimensional spaces, the performance can degrade toward a linear scan over the whole database [6]. So, the challenge is on these spaces, where the distance histogram becomes very concentrated; in other words, all the pair-wise distances between elements are almost equal.

A practical approach for high dimensional spaces is to resign to do not obtain an exact similarity answer. Instead, we can settle for approximate answers, that means that we can miss some relevant elements from the query answer set, or report some non relevant objects [16]. In this scenario, the goal is to design an efficient method whose answer quality is within certain bounds.

In [3,4], the authors present a novel approximate method for similarity searching in metric spaces, which was coined the *permutation based algorithms* (PBA). Experimentally, this method has shown an extremely competitive performance in high dimensional spaces [15]. However, its performance can still be improved with a different dissimilarity measure between permutations, as we show in this work.

The rest of this paper is organized as follows. Section 2 reviews the PBA's basic concepts and some related work. We introduce our proposal in Sect. 3, and show its experimental evaluation in Sect. 4. Finally, we give some conclusion remarks and mention some lines of future work in Sect. 5.

2   Basic Concepts

A metric space is a pair $(\mathbb{X}, d)$, where $\mathbb{X}$ is the universe of valid objects with a distance function $d : \mathbb{X} \times \mathbb{X} \rightarrow \mathbb{R}^+$. The distance d satisfies the following properties: symmetry $d(x, y) = d(y, x)$, reflexivity $d(x, x) = 0$, strict positiveness $d(x, y) > 0 \iff x \neq y$, and triangle inequality $d(x, y) \leq d(x, z) + d(z, y)$, $\forall\, x, y, z \in \mathbb{X}$. The database is a finite subset of the valid objects $\mathbb{U} \subseteq \mathbb{X}$.

Let $q \in \mathbb{X}$ be a given query object, $k \in \mathbb{N}$, and $r \in \mathbb{R}^+$. Formally, the basic kinds of similarity queries are: (i) the $k-nearest$ $neighbors$ $query$ is a set $S = \{u_1, \ldots, u_k\}$, $|S| = k$, $S \subseteq \mathbb{U}$, such that $\forall\, u_i \in S, v \in \mathbb{U} \setminus S, d(q, u_i) \leq d(q, v)$; and (ii) the *range query* is a set $S = \{u \in \mathbb{U}, d(q, u) \leq r\}$.

The general approach to proximity searching in metric spaces considers two stages, namely, preprocessing and query response. During the preprocessing stage, an index is built in order to efficiently get answers in the query stage.

Currently, the state-of-the-art methods are divided in three big families [4, 6,13,17,20], namely, *pivot based algorithms*, *compact partition algorithms*, and *permutation based algorithms*. A *pivot based algorithm* chooses a small set of elements called *pivots*, and every pivot computes and stores all the distances to the rest of the elements. At query time, distant elements are discarded using the triangle inequality and the stored distances. A *compact partition algorithm* splits the space using some reference objects, called *centers*. The centers define partitions, and the objects are placed in one partition. At query time, partitions that do not intersect with the query are discarded along with their inner objects.

Since pivot based and compact partition algorithms do not work well enough in high dimension spaces, both are out of our scope. So, we only describe in detail the permutation based algorithms.

2.1   Permutation Based Algorithms

In [3,4], the authors define a subset of *permutants* as $\mathbb{P} = \{p_1, \ldots, p_m\}$, where $\mathbb{P} \subseteq \mathbb{U}$. The simplest way to choose permutants is at random, however, there are better alternatives [1]. Each element of the space $u \in \mathbb{U}$ defines a permutation Π_u by ordering the permutants according to the distances to them. That is, for all $1 \leq i < m$, it satisfies $d(\Pi_u(i), u) \leq d(\Pi_u(i+1), u)$. We use $\Pi_u^{-1}(i)$ for the position of a permutant p_i in Π_u. The resulting set of permutations conforms the index. The work hypothesis of this technique is *two equals elements must have the same permutation, while two similar elements will, hopefully, have similar permutations.*

Up to now, authors have proposed to measure the similarity between two permutations using *Spearman Rho* metric [4,7,9] defined as follows:

$$S_\rho(\Pi_u, \Pi_q) = \sqrt{\sum_{1 \leq i \leq m} (\Pi_u^{-1}(i) - \Pi_q^{-1}(i))^2} \tag{1}$$

As and example, let $\Pi_q = [4, 6, 5, 3, 1, 2]$ and $\Pi_u = [2, 1, 4, 3, 6, 5]$ be the permutation of the query and the one of an element u, respectively. It can be noticed that the particular element p_1 in permutation Π_u is found three positions before its position in Π_q; that is, $\Pi_u^{-1}(1) = 2$ and $\Pi_q^{-1}(1) = 5$. The differences between permutations are $(2 - 5), (1 - 6), (4 - 4), (3 - 1), (6 - 3), (5 - 2)$. Thus, the sum of their squares is 56, and $S_\rho(\Pi_u, \Pi_q) = \sqrt{56}$.

There are other similarity measures between permutations [9], such as *Spearman Footrule* and *Kendall Tau*. They were first used and described in [3], when the Permutation-based method was introduced. The definition of Spearman Footrule metric is as follows:

$$S_F(\Pi_u, \Pi_q) = \sum_{1 \leq i \leq m} |\Pi_u^{-1}(i) - \Pi_q^{-1}(i)| \tag{2}$$

Kendall Tau counts the number of inversions needed to convert one permutation into the other. Formally, for every pair $p_i, p_j \in \mathbb{P}$, if p_i and p_j are in the same order in Π_u and Π_q (that is, $\Pi_u^{-1}(i) < \Pi_u^{-1}(j) \iff \Pi_q^{-1}(i) < \Pi_q^{-1}(j)$) then $K_{p_i,p_j}(\Pi_u, \Pi_q) = 0$; otherwise $K_{p_i,p_j}(\Pi_u, \Pi_q) = 1$. So, Kendall Tau is:

$$K(\Pi_u, \Pi_q) = \sum_{p_i,p_j \in \mathbb{P}} K_{p_i,p_j}(\Pi_u, \Pi_q) \tag{3}$$

Then, given a query $q \in \mathbb{X}$, we compute Π_q and compare it with all the permutations stored in the index. Next, $\mathbb{U}$ is traversed in increasing permutation dissimilarity order, comparing directly the objects in $\mathbb{U}$ with the query using the distance d of the particular metric space. If we limit the number of distance computations, we obtain an approximate search algorithm, that is able to find query answer within certain quality bounds.

Note that the distance computations performed for computing Π_q are compulsory if we use the PBA for solving a similarity query. In the literature, these distance computations, which are mandatorily performed in case we use some index, are called the *internal comparisons*. On the other hand, the ones needed for traversing the objects in $\mathbb{U}$ are called the *external comparisons* [6]. Usually, given a particular index, the number of internal comparisons is fixed, while the number of external ones depends on the particular query.

Related Work. Other existing proposals about permutation based algorithms use the same metrics between permutations (i.e. Spearman Footrule or Spearman Rho) during the query stage. For example:

- In [2], the Metric Inverted File is presented. Its construction is the same as Permutation-based algorithm, but only a prefix of the first $m' \leq m$ values in each permutation are used for indexing. For the query q, they use the first $m_q \leq m'$ permutation elements. Authors proposed to store this index in an inverted file, so that, for each permutant they store a posting list that contains the object and the position of the permutant in its respective object permutation. They also use the inverted index for avoiding sequential scan. Once a query is given, they compute Spearman Footrule. When this is not possible because a permutant is not in the prefix of m_q permutants of the query permutation, they used the m_q value to replace it. An a variant proposed in [19] also used Spearman Footrule.
- In [8], permutations are treated as strings. The authors used the closest $l \leq m$ permutants and called it the *Permutation Prefix Index*. This way, the permutation prefixes are stored in a suffix tree index. At query time, the use Spearman Footrule metric while traversing the suffix tree branches, pruning the alternatives whenever is possible.
- In [14], the authors propose a variant of the PBA that also uses the prefixes, but in a quantized fashion. Besides, they discuss several computational machinery to manage the index both in RAM and secondary memory, with several fixes to improve the efficiency in the computations and storage.
- There are other proposed methods to merge some known indexes with PBA, for instance, the *List of Clustered Permutations* [10] or *Fixed Height Queries Tree of Permutations* [12]. However, they also use Spearman Footrule.

3   Our Proposal

Until now, each permutation based algorithm uses either Spearman Footrule or Spearman Rho metrics in order to measure the similarity between permutations. Our proposal consists in providing a new family of dissimilarity measures between permutations for proximity searching.

We have observed that given two permutations, the fact that a single permutant appears in far away positions in these permutations is enough to assume that those objects are far away according to the real distance of the metric space. Furthermore, the permutants that have medium size variations in their positions can introduce noise in the assessment of the similarity between permutations. These observations lead us to two simple ideas, namely: (i) we should amplify the big permutant position differences when assessing the dissimilarity between permutations, and (ii) we should avoid the use of the medium size differences.

For this purpose, we consider that each permutation is split in three sections. So, a permutation Π_u of an element $u \in \mathbb{U}$ is partitioned in: permutants near to u (we named this part as β), permutants at the middle of Π_u (we called this part as γ), and finally those permutants far to u (this part is named as δ). Note that the permutants in the γ section have higher chance of producing a medium size position differences when computing the dissimilarity between two permutations, while the ones in sections β and δ can produce great differences in positions.

For example, if we consider the same Π_q aforementioned, its partitioning would be:

$$\Pi_q = [\ \underbrace{4,6}_{\beta}, \underbrace{5,3}_{\gamma}, \underbrace{1,2}_{\delta}\ ] \tag{4}$$

We define ϕ_i, $1 \le i \le m$, for some Π_u and Π_q, as follows:

$$\phi_i = |\Pi_u^{-1}(i) - \Pi_q^{-1}(i)|, \quad 1 \le i \le m \tag{5}$$

so, we can redefine Spearman Footrule (S_F) and Spearman Rho (S_ρ) metrics using ϕ_i:

$$S_F(\Pi_u, \Pi_q) = \sum_{i=1}^{m} \phi_i$$

$$S_\rho(\Pi_u, \Pi_q) = \sqrt{\sum_{i=1}^{m} \phi_i^2}$$

It can be noticed that the bigger the value of ϕ_i, the greater is its impact on the position of an element u in the review ordering of $\mathbb{U}$. Therefore, our proposal conveniently amplifies some values of ϕ_i to obtain a better order, induced by the dissimilarity between permutations, to review the elements in $\mathbb{U}$ at search time.

Thus, we consider to amplify those values of ϕ_i which are larger than a selected threshold μ by an amplification factor of $\alpha \ge 1$. Hence, our new family of dissimilarity measures between permutations introduces the parameter α, which is responsible to amplify the large inversions of permutants; and the interval I, that indicates if we consider the whole permutation ($I = \{1, \ldots, m\}$) or only its first and last third, called β and δ, so $I = \beta \circ \delta$, excluding the middle third γ. We can use a new value Φ_i, which depends on whether ϕ_i surpasses or not a threshold μ. Therefore, our new family $T_{I,\alpha}(\Pi_u, \Pi_q)$ is defined as:

$$T_{I,\alpha}(\Pi_u, \Pi_q) = \sum_{i \in I} \Phi_i \tag{6}$$

where

$$\Phi_i = \begin{cases} \phi_i^\alpha & : \phi_i \geq \mu \\ \phi_i & : \phi_i < \mu \end{cases} \tag{7}$$

In order to abreviate the notation, we use C (for Complete) to indicate that $I = \{1, \ldots, m\}$ and P (for Partial) to name the option where $I = \beta \circ \delta$.

In order to show the goodness of this family, in the experimental evaluation we show the following cases: $T_{C,2}, T_{P,2}, T_{C,\log_{10}(\phi_i)}$, and $T_{P,\log_{10}(\phi_i)}$.

It is necessary to emphasize that although the value of α seems to be independent of the threshold μ, in order to achieve measures that satisfy the above criteria, we always use a value of μ that ensures $\log_{10}(\phi_i) \geq 1$.

We remark that even though the dissimilarity measures belonging to the $T_{I,\alpha}$ family are not metrics, as they do not satisfy the triangle inequality, they do induce a competitive reviewing order of the set $\mathbb{U}$, which is the actual objective of the permutation dissimilarity measure.

$T_{I,\alpha}$ **is Semimetric.** To prove that $T_{I,\alpha}$ is semimetric we need to show that it satisfies the following properties: symmetry, reflexivity and non-negativity [18].

- Symmetry: $T_{I,\alpha}(\Pi_x, \Pi_y) = T_{I,\alpha}(\Pi_y, \Pi_x)$, since they are the sum of several Φ_i. Each Φ_i can be the absolute value of a difference, or a power $\alpha \geq 1$ of that absolute value. As the absolute value of a difference is commutative, so the sum must be the same.
- Reflexivity: $T_{I,\alpha}(\Pi_x, \Pi_x) = 0$, because any $\phi_i = |\Pi_x^{-1}(i) - \Pi_x^{-1}(i)| = 0$.
- Non-negativity: $T_{I,\alpha}(\Pi_x, \Pi_y) \geq 0$, because is the sum of several values that are greater than or equal to zero.

4   Experimental Evaluation

In this section, we show the performance of our proposal over some real-life databases, available at SISAP Metric Library benchmark set [11]. For each dataset we randomly choose 500 objects as queries and we index the rest of the elements. We tested $T_{I,\alpha}$ for several combinations of α and I, and now we show four of them, with competitive performance, namely, $T_{C,2}, T_{P,2}, T_{C,\log_{10}(\phi_i)}$, and $T_{P,\log_{10}(\phi_i)}$. We have considered different values for μ, and we show the ones with the best behavior. In the plots, we only show the computed distances when traversing $\mathbb{U}$ in increasing permutation dissimilarity order; these are also know as the number of external comparisons. We do not show the internal comparisons as they are all the same for the tested PBA alternatives considered in this experimental evaluation.

4.1   Colors

This database consists of 112,682 color histograms of images, represented as 112-dimensional vectors. As any quadratic form can used to compare objects, we use Euclidean distance as the simplest alternative.

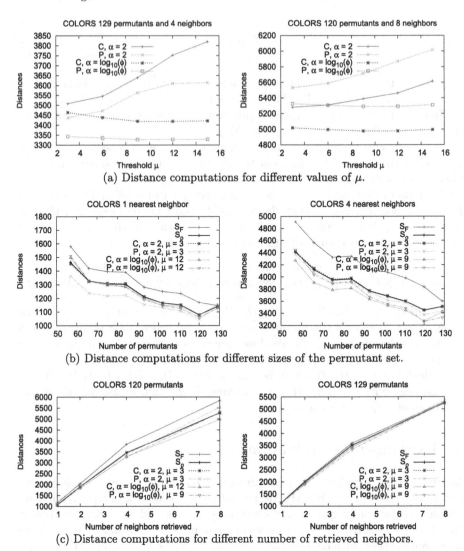

(a) Distance computations for different values of μ.

(b) Distance computations for different sizes of the permutant set.

(c) Distance computations for different number of retrieved neighbors.

Fig. 1. Performance of our proposal using COLORS database.

In Fig. 1, we show the experimental comparison of PBA using different dissimilarity measures between permutations in order to compute the reviewing order of $\mathbb{U}$, varying all the parameters under study.

We start by studying the effect of μ, see Fig. 1(a), when retrieving four and eight nearest neighbors per query and two sizes of the permutant set. We show the most significant results. As can be seen, when we use $\alpha = 2$, the best results are obtained with $\mu = 3$, independently of considering complete or partial interval ($I = C$ or $I = P$). On the other hand, when we consider $\alpha = \log_{10} \phi$, the best performance is obtained with $\mu = 9$ or $\mu = 12$.

The second parameter under analysis is the size of the permutant set. In this test, we use the selected values of μ, for several sizes of the answer set. Figure 1(b) shows that as long as the number of permutants increases, the performance of the technique improves up to 120 permutants, and the second best size is 129 in most cases.

Finally, in Fig. 1(c), we show the results when varying the size of the answer set using the best combinations of the parameters μ and permutant set size. As it can be seen, the PBA method using Spearman Footrule and Spearman Rho metrics can be overcome. In this particular database, the best performance is obtained when we use $T_{C,\log_{10}(\phi_i)}$ and $T_{P,\log_{10}(\phi_i)}$, and we can reduce up to 5% of the distance computations.

It is important to mention that when we use $T_{P,\alpha}$, the index save the space of one third of the permutants, so the needed space to store the index is just two thirds of the one used by the other alternatives.

4.2   NASA

This dataset consists of 40,150 feature vectors in $\mathbb{R}^{20}$. These 20-dimensional vectors were generated from images downloaded from NASA (available at http://www.dimacs.rutgers.edu/Challenges/Sixth/software.html), where duplicate vectors were eliminated. We used the Euclidean distance to compare the feature vectors of this collection of images.

Figure 2 shows the experimental comparison of PBA using different dissimilarity measures between permutations, varying all the parameters under study.

Figure 2(a) shows the effect of μ when retrieving four nearest neighbors per query, and two sizes of the permutant set. As can be seen, when we use $\alpha = 2$, the best performance is obtained with $\mu = 3$, both for complete or partial interval ($I = C$ or $I = P$). On the other hand, when we consider $\alpha = \log_{10}(\phi_i)$, there is no value of μ that shows a significant advantage.

With respect to the size of the permutant set, Fig. 2(b) shows that when solving k–NN queries, the best performance for all the alternatives is obtained with 120 or 129 permutants. However, when using $\alpha = \log_{10}(\phi_i)$ in 1–NN queries, we also appreciate another local optimum with 102 permutants.

Finally, in Fig. 2(c), we show that the alternative $T_{C,\log_{10}(\phi_i)}$ surpasses the PBA with traditional Spearman Footrule or Spearman Rho metrics.

4.3   Dictionaries

Another real-life databases are a Spanish dictionary with 89,000 words and an English one with 69,069 words. In both cases, we use the edit distance to measure the similarity between words. The edit distance counts the number of character insertions, deletions or substitutions needed to transform a word into the other.

In Fig. 3, we show the performance of our proposal over both dictionaries. For economy of space, we only show the effects of varying the size of the permutant set and the size of the query answer, with the best values of μ according to our

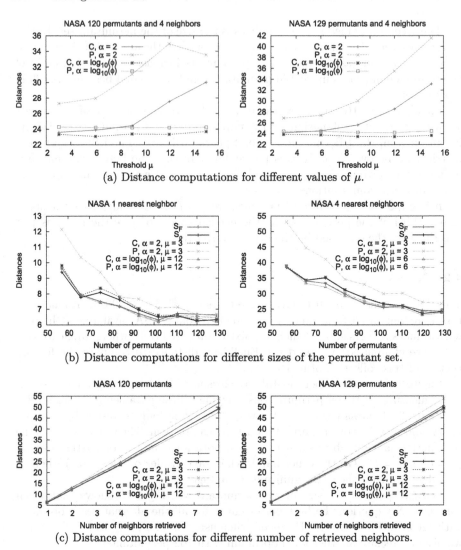

(a) Distance computations for different values of μ.

(b) Distance computations for different sizes of the permutant set.

(c) Distance computations for different number of retrieved neighbors.

Fig. 2. Performance of our proposal using NASA database.

experimental evaluation. We report results for the Spanish dictionary on the left column, and the ones for English dictionary on the right column.

Figures 3(a) and (b) shows that with 129 permutants we obtain a competitive performance for both 4–NN and 8–NN queries, and the two considered dictionaries. There is another local minimum with 111 permutants, which is also the best performance for S_ρ, but it is surpassed by our proposal with 129 permutants.

Figure 3(c) shows that the worst alternative is to use Spearman Footrule, and also shows that all the studied alternatives of our proposal beat both traditional Spearman metrics. Considering that the best traditional dissimilarity measure

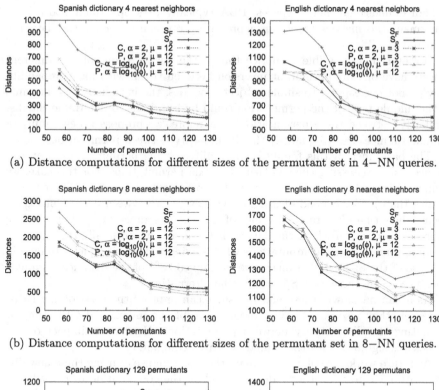

(a) Distance computations for different sizes of the permutant set in 4−NN queries.

(b) Distance computations for different sizes of the permutant set in 8−NN queries.

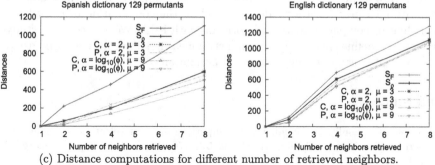

(c) Distance computations for different number of retrieved neighbors.

Fig. 3. Performance of our proposal using Spanish and English dictionaries.

is Spearman Rho, we remark that our proposal can save up to 15% of distance computations regarding Spearman Rho performance in the case of the English dictionary. Furthermore, the saving goes up to 30% of distance computations in the case of the Spanish dictionary.

5   Conclusions

The *Permutation Based Algorithm* (PBA) is a novel technique for approximate similarity searching. Experimentally, the PBA has shown to be very successful in high dimensional metric spaces. PBA typically uses Spearman Footrule or Spearman Rho metrics to obtain a promising revision order for the database elements.

However, it is possible to notice that if the permutations of two different objects have one big permutant inversion between them, with high probability they do not correspond to similar objects. Besides, the inversion of permutants in the middle part of the permutation could be irrelevant to determine the dissimilarity between these permutations. Therefore, we propose a new family of semimetric measures to quantify the dissimilarity between permutations.

The basic idea of this family is to amplify the big permutant position differences when assessing the similarity between permutations. For this sake, we define a threshold μ, and we amplify the permutation position differences ϕ greater than μ. So, instead of using ϕ, whenever $\phi \geq \mu$, we use ϕ^α. Our experiments show that the practical values of α are in the range $(1, 2]$. This way, we improve the order in which the dataset is traversed, which finally translates on avoiding some distance computations when solving a similarity query. In fact, in the case of the Spanish dictionary, we save up to 30% of the distance computations needed by the classic Spearman Rho alternative.

As it can be shown, these new measures can improve the costs of any PBA. Moreover, they also could save storage space of the index, and CPU time at evaluating the dissimilarity between permutations. Hence, the impact of our proposal is significant for the similarity search community.

As future works, we plan to evaluate the performance of using these new dissimilarity measures to other indexes, based on the combination of pivot-based algorithms or compact partitions algorithms with PBA. We also consider studying how to redefine this family of semimetric measures if we have a set of groups of permutants instead of only one set of permutants.

References

1. Amato, G., Esuli, A., Falchi, F.: A comparison of pivot selection techniques for permutation-based indexing. Inf. Syst. **52**, 176–188 (2015). https://doi.org/10.1016/j.is.2015.01.010
2. Amato, G., Savino, P.: Approximate similarity search in metric spaces using inverted files. In: 3rd International ICST Conference on Scalable Information Systems, INFOSCALE 2008, Vico Equense, Italy, 4–6 June 2008. p. 28. ICST/ACM (2008). https://doi.org/10.4108/ICST.INFOSCALE2008.3486
3. Chávez, E., Figueroa, K., Navarro, G.: Proximity searching in high dimensional spaces with a proximity preserving order. In: Gelbukh, A., de Albornoz, Á., Terashima-Marín, H. (eds.) MICAI 2005. LNCS (LNAI), vol. 3789, pp. 405–414. Springer, Heidelberg (2005). https://doi.org/10.1007/11579427_41
4. Chávez, E., Figueroa, K., Navarro, G.: Effective proximity retrieval by ordering permutations. IEEE Trans. Pattern Anal. Mach. Intell. (TPAMI) **30**(9), 1647–1658 (2009)
5. Chávez, E., Navarro, G.: A probabilistic spell for the curse of dimensionality. In: Buchsbaum, A.L., Snoeyink, J. (eds.) ALENEX 2001. LNCS, vol. 2153, pp. 147–160. Springer, Heidelberg (2001). https://doi.org/10.1007/3-540-44808-X_12

6. Chávez, E., Navarro, G., Baeza-Yates, R., Marroquín, J.: Proximity searching in metric spaces. ACM Comput. Surv. **33**(3), 273–321 (2001)
7. Diaconis, P., Graham, R.L.: Spearman's footrule as a measure of disarray. J. R. Stat. Soc. Ser. B (Methodol.) **39**(2), 262–268 (1977)
8. Esuli, A.: Use of permutation prefixes for efficient and scalable approximate similarity search. Inf. Process. Manage. **48**(5), 889–902 (2012). https://doi.org/10. 1016/j.ipm.2010.11.011
9. Fagin, R., Kumar, R., Sivakumar, D.: Comparing top k lists. SIAM J. Discrete Math. **17**(1), 134–160 (2003)
10. Figueroa, K., Paredes, R.: List of clustered permutations for proximity searching. In: Brisaboa, N., Pedreira, O., Zezula, P. (eds.) SISAP 2013. LNCS, vol. 8199, pp. 50–58. Springer, Heidelberg (2013). https://doi.org/10.1007/978-3-642-41062-8_6
11. Figueroa, K., Navarro, G., Chávez, E.: Metric spaces library (2007). http://www. sisap.org/Metric_Space_Library.html
12. Figueroa, K., Paredes, R., Camarena-Ibarrola, J.A., Reyes, N.: Fixed height queries tree permutation index for proximity searching. In: Carrasco-Ochoa, J.A., Martínez-Trinidad, J.F., Olvera-López, J.A. (eds.) MCPR 2017. LNCS, vol. 10267, pp. 74–83. Springer, Cham (2017). https://doi.org/10.1007/978-3-319-59226-8_8
13. Hjaltason, G., Samet, H.: Index-driven similarity search in metric spaces. ACM Trans. Database Syst. **28**(4), 517–580 (2003). https://doi.org/10.1145/958942. 958948
14. Mohamed, H., Marchand-Maillet, S.: Quantized ranking for permutation-based indexing. Inf. Syst. **52**, 163–175 (2015). https://doi.org/10.1016/j.is.2015.01.009
15. Naidan, B., Boytsov, L., Nyberg, E.: Permutation search methods are efficient, yet faster search is possible. Proc. VLDB Endow. **8**(12), 1618–1629 (2015). https:// doi.org/10.14778/2824032.2824059
16. Patella, M., Ciaccia, P.: Approximate similarity search: a multi-faceted problem. J. Discret Algorithms **7**(1), 36–48 (2009). https://doi.org/10.1016/j.jda.2008.09. 014. Selected papers from the 1st International Workshop on Similarity Search and Applications (SISAP)
17. Samet, H.: Foundations of Multidimensional and Metric Data Structures (The Morgan Kaufmann Series in Computer Graphics and Geometric Modeling). Morgan Kaufmann Publishers Inc., San Francisco (2005)
18. Skopal, T.: On fast non-metric similarity search by metric access methods. In: Ioannidi, Y., et al. (eds.) EDBT 2006. LNCS, vol. 3896, pp. 718–736. Springer, Heidelberg (2006). https://doi.org/10.1007/11687238_43
19. Tellez, E.S., Chavez, E., Navarro, G.: Succint nearest neighbor search. Inf. Syst. **38**(7), 1019–1030 (2013)
20. Zezula, P., Amato, G., Dohnal, V., Batko, M.: Similarity Search: The Metric Space Approach, Advances in Database Systems, vol. 32. Springer, Heidelberg (2006). https://doi.org/10.1007/0-387-29151-2

LID-Fingerprint: A Local Intrinsic Dimensionality-Based Fingerprinting Method

Michael E. Houle[1], Vincent Oria[2], Kurt R. Rohloff[2], and Arwa M. Wali[2,3(✉)]

[1] National Institute of Informatics, Tokyo 101-8430, Japan
meh@nii.ac.jp
[2] New Jersey Institute of Technology, Newark, NJ 07102, USA
{vincent.oria,kurt.rohloff,amw7}@njit.edu
[3] King Abdulaziz University, Jeddah, Saudi Arabia

Abstract. One of the most important information hiding techniques is fingerprinting, which aims to generate new representations for data that are significantly more compact than the original. Fingerprinting is a promising technique for secure and efficient similarity search for multimedia data on the cloud. In this paper, we propose *LID-Fingerprint*, a simple binary fingerprinting technique for high-dimensional data. The binary fingerprints are derived from sparse representations of the data objects, which are generated using a feature selection criterion, Support-Weighted Intrinsic Dimensionality (support-weighted ID), within a similarity graph construction method, NNWID-Descent. The sparsification process employed by LID-Fingerprint significantly reduces the information content of the data, thus ensuring data suppression and data masking. Experimental results show that LID-Fingerprint is able to generate compact binary fingerprints while allowing a reasonable level of search accuracy.

Keywords: Intrinsic dimensionality · *K*-nearest neighbor graph
Fingerprinting · Information hiding

1 Introduction

The increasing amount of private multimedia data (documents, images, audio and videos) stored on the cloud presents new challenges in regard to secure and efficient search. Cloud storage is subject to such abuses as 'Man-in-the-Cloud' (MITC) attacks, data breaches and data loss, and phishing attacks. Passive adversary ('honest-but-curious') attack, in which information is inferred without explicit data theft, is also a constant threat.

Information hiding is the branch of information security branches concerned with the obscuring from unintended observers of the search and exchange of information [23]. Topics related to information hiding include covert channels [24], steganography [29], anonymity [16], and watermarking [8].

© Springer Nature Switzerland AG 2018
S. Marchand-Maillet et al. (Eds.): SISAP 2018, LNCS 11223, pp. 134–147, 2018.
https://doi.org/10.1007/978-3-030-02224-2_11

Information hiding can be accomplished through the creation of a compact and unique identifier for each data object. This process, referred to as *fingerprinting*, can be active or passive. In active fingerprinting, the unique identifier is embedded into a digital object using watermarking techniques. Passive fingerprinting, which is the focus of this paper, uses the original features of an object constructs an identifier (digital signature) for it [23]. In addition to providing copyright and data privacy, fingerprinting can be used to trace illegal usage of the data [3], or to verify the integrity of the data [6]. Typically, fingerprints have much shorter and compact representations than the actual objects, and are commonly expressed as binary codes (bit vectors). Short, compact representations are desirable in that they lead to better efficiency in search. However, care must be taken to ensure that the representations are unique.

There has been much research investigating the generation of compact binary codes in a variety of digital data domains. State-of-the-art methods based on Locality Sensitive Hashing (LSH) [2,7] have been successfully used to generate audio and image fingerprints [28,32]. Machine learning techniques [5,31,32,34], Spectral Hashing (SH) [35], bag-of-words approaches [11], and triplet histograms [10] have also been considered for fingerprinting purposes.

In this paper, we propose *LID-Fingerprint*, a simple binary fingerprinting technique for high-dimensional data. LID-Fingerprint can be used for hiding information on the server side (cloud) as a way of preventing passive adversarial attack. The binary fingerprints are derived from sparse representations of the data objects through the NNWID-Descent similarity graph construction process [20], in which local feature sparsification steps are interleaved with neighborhood refinement steps. For each object, NNWID-Descent identifies features to be sparsified according to a feature selection criterion, Support-Weighted Intrinsic Dimensionality (support-weighted ID) [18], which assesses the contribution of each feature to the overall intrinsic dimensionality, and measures the ability of the feature to discriminate between nearby objects. Those features for which the local discriminative power is lowest are set to 0, and the surviving features are set to 1. The sparsification process employed by LID-Fingerprint significantly reduces the information content of the data, thus ensuring two of the standard criteria for protecting data privacy: data suppression and data masking. Experimental results show that LID-Fingerprint is able to generate compact binary fingerprints while allowing a reasonable level of search accuracy.

The remainder of this paper is organized as follows. Section 2 provides background on the relevant fingerprinting research literature. An overview of the NNWID-Descent framework is presented in Sect. 3. We discuss the proposed LID-Fingerprint technique in Sect. 4. In Sect. 5, the performance of our method, along with the experimental results and analysis using several real datasets, is compared to that of NNWID-Descent and competing methods from the literature. Finally, we conclude in Sect. 6 with a discussion of future research directions.

2 Fingerprint Generation

A number of works on the design and generation of fingerprints have been proposed in the literature. In addition to the withholding of information from unauthorized observers on network servers or the cloud, the common objective of these works is the acceleration of nearest neighbor search. Locality Sensitive Hashing (LSH) [14, 30]—perhaps the state-of-the-art fingerprint generation method—seeks to find an efficient binary representations of high-dimensional data objects by computing hash functions based on random projections. Each random projection contributes one or more bits to the object fingerprint. The hash functions help in maintaining the similarity between objects in the new binary space [34]. Other LSH-based binary fingerprinting methods have also been proposed, such as MinHash [4], SuperBit [22], Simhash [7], and Spectral Hashing [35]. Although LSH and its variants work efficiently for high-dimensional datasets, it has been reported that when the number of bits is fixed and relatively small, LSH may perform rather poorly as regards the accuracy of similarity search on the generated fingerprints [34].

A common step for many fingerprinting methods is to include dimensionality reduction techniques within binary code generation. For example, in [34] Torralba et al. converted image GIST descriptors into compact binary codes by adapting machine learning techniques for dimensionality reduction, such as Boosting [32] and Restricted Boltzmann Machines [31], for use with LSH. This method allows fast object recognition with an accuracy level comparable with that of full descriptors. Strecha et al. [33] applied Linear Discriminant Analysis (LDA) as a dimensionality reduction technique on the original SIFT vectors. Binary codes are then learned by finding a global matrix projection using the gradient-based AdaBoost method [27]. Caballero et al. [5] proposed FiG, an automatic active fingerprint generation system. Their system automatically generates candidate queries, sends them to a set of training hosts, identifies useful queries, and applies machine learning techniques (including dimensionality reduction) to generate sets of possible fingerprints.

The aforementioned methods are supervised, in that they depend on object label information in order to generate binary codes. Unsupervised fingerprinting methods have also been developed. In [36], two quantization methods, Spectral Bits and Phase Bits, were proposed for converting real-valued spectral minutiae features into binary codes. The authors applied two feature reduction techniques on the spectral minutiae features, Column Principle Component Analysis (CPCA) and Line Discrete Fourier Transform (LDFT) [37], before using quantization to extract compact fixed-length binary fingerprint templates. The proposed methods mask out (set to 0) those features having absolute values below certain thresholds, while setting the remaining features to 1. Gong and Lazebnik [15] defined an iterative quantization (ITQ) method for generating binary codes. The method starts by transforming the data using a PCA-based binary coding scheme, followed by alternating minimization refinement on these transformed data in order to reduce the quantization error.

Farooq et al. [10] presented an anonymous fingerprinting construction technique that satisfies two criteria, anonymity and recoverability. In the first phase, invariant features are selected from the original fingerprint, and then converted to binary representations using template-based matching based on minutiae triplets. These binary representations are then transformed into anonymous representations by randomizing the user template by means of a uniquely-assigned key, which can be redefined if the template were ever to become compromised. However, this anonymization strategy is computationally very expensive, as it must precompute all possible triples of invariant features.

3   NNWID-Descent

As the basis for the work presented in this paper, in this section we provide a brief review of the NNWID-Descent similarity graph construction framework. We describe its dimensionality reduction criterion, Support-Weighted Intrinsic Dimensionality (support-weighted ID), and discuss its utilization in the feature ranking and sparsification processes.

3.1   Support-Weighted Intrinsic Dimensionality Measure

Support-Weighted Intrinsic Dimensionality (support-weighted ID, or wID) [20] is an extension of the Local Intrinsic Dimensionality (LID) measure introduced in [1,19]. Given a distribution of distances with a univariate cumulative distribution function ϕ that is positive and continuously differentiable in the vicinity of distance value $r > 0$, the indiscriminability of ϕ at r is given by

$$\mathrm{ID}_\phi(r) \triangleq \frac{r \cdot \phi'(r)}{\phi(r)}. \tag{1}$$

The indiscriminability reflects the growth rate of the cumulative distance function at r; it can be regarded as a probability density associated with the neighborhood of radius r (that is, $\phi'(r)$), normalized by the cumulative density of the neighborhood (that is, $\phi(r)/r$). The local intrinsic dimension has been shown to be equivalent to a notion of local intrinsic dimensionality, which can be defined as the limit $\mathrm{ID}_\phi^* = \lim_{r \to 0^+} \mathrm{ID}_\phi(r)$. In general, estimation of ID_ϕ^* requires samples that are the result of a k-nearest neighbor query on the underlying dataset. For the experiments of this paper, we make use of the maximum likelihood estimator presented in [1].

Although ID_ϕ adequately models the dimensional characteristics within a given locality, it does not in itself account for the relative importance of different localities. One way to adjust for relative importance is to give a weighting proportional to the probability measure associated with the locality. In [20], the support-weighted ID measure

$$\mathrm{wID}_\phi(r) \triangleq \phi(r) \cdot \mathrm{ID}_\phi(r) = r \cdot \phi'(r)$$

was shown to have interesting decomposition properties when the distance r was determined across a collection of features, allowing for the comparison of the discriminabilities of different features in a dataset. For more details regarding to the support-weighted ID model, we refer the readers to [17,20].

In practice, for reasonably large dataset sizes, the following approximation can be used for the support-weighted ID measure [17,18]:

$$\mathrm{wID}_\phi(r) \approx \frac{k\,\mathrm{ID}^*_\phi}{n} \cdot \left(\frac{r}{w}\right)^{\mathrm{ID}^*_\phi}, \tag{2}$$

where n is the size of the dataset, and w is the distance to the k-th nearest neighbor of a given sample point.

3.2   Feature Ranking and Sparsification Based on wID

Here, we show how support-weighted ID was used in [18] for feature ranking with respect to individual data objects. Assume that we have a dataset X with n data objects represented as feature vectors in $\mathbb{R}^D$. The set of features is denoted as $F = \{1, 2, \ldots, D\}$, such that $j \in F$ is the j-th feature in the vector representation. For each object-feature combination, we consider a neighborhood of size $K \geq k$, determined using only the single feature in question. We further assume that the vectors are normalized. Then, since the factor k/n in Eq. 2 can be regarded as constant, for an object $x_i \in X$, the support-weighted ID score for each feature f_j was simplified using the following formula [18]:

$$\mathrm{wID}_i(f_j) = \mathrm{ID}^*_{f_j} \cdot \left(\frac{a}{w_{f_j}}\right)^{\mathrm{ID}^*_{f_j}}, \tag{3}$$

where $\mathrm{ID}^*_{f_j}$ is the LID estimate for the neighborhood with respect to feature f_j, and w_{f_j} is the distance to the K-th nearest neighbor with respect to feature f_j. In practice, the value a represents a small positive distance (r in Eq. 2). For simplicity, a can be set as an average across many objects of a sample of K-NN distances with respect to feature f_j.

Equation (3) helps to find the most discriminative features by considering both the density of neighborhood around each object and the complexity of local ID with respect to a particular feature f_j. In NNWID-Descent, the most discriminative features for each object are ranked in descending order of $\mathrm{wID}_i(f_j)$; from these, a proportion Z of the top-ranked features are deemed to be noise and designated as candidates for sparsification. In the sparsification process, the impact of noisy features is minimized by setting their values to the global mean, which (due to normalization) is zero.

3.3   NNWID-Descent

The NNWID-Descent framework interleaves the feature ranking and sparsification process with k-NN graph construction using NN-Descent [9], where k is the target neighborhood size for each object in the output graph G. Algorithm 1 gives the complete algorithm for NNWID-Descent. The algorithm has two phases: an initialization phase, and a sparsification and refinement phase. In the initialization phase, after normalizing the original vectors of the dataset X, the algorithm computes an initial approximate k-NN graph using NN-Descent [9] (lines 2–3). The NN-Descent procedure depends on the so-called *local join* operation—given a target point x, the local join operation checks whether any neighbor of x's neighbors is closer to x than any points currently in its neighbor list, and also whether pairs of neighbors of x can likewise improve each other's tentative neighbor list. In the sparsification and refinement phase (lines 6–16), feature ranking uses wID to identify those features that can most safely be declared as noise, and then sparsifies a target proportion of them. After the sparsification process, the k-NN graph entries are improved through refinement.

Algorithm 1: NNWID-Descent.

Input : Dataset X, distance function dist, neighborhood size k for the graph, neighborhood size K for computing wID scores, sparsification rate Z, number of iterations T.

Output: k-NN graph G.

1 {Initialization Phase}
2 Normalize the original feature vectors of X;
3 Run NN-Descent(X, dist, k) to convergence to obtain an initial k-NN graph G;
4 Set the value of a to the average of k-NN distances computed for the feature over a sample of objects;
5 {Sparsification and Refinement Phase}
6 **repeat**
7 | Generate a list L of all data points of X in random order;
8 | **foreach** *data point $x \in L$* **do**
9 | For each feature, compute the wID (Equation 3) using the K-NN distances of x for this feature with respect to the current set of feature vectors of X;
10 | Rank the features of x in descending order of their wID scores;
11 | Change the value of the top-ranked Z-proportion of features to 0;
12 | Recompute the distances from x to its k-NN and RNN points;
13 | Re-sort the k-NN lists of x and its RNNs;
14 | For each pair (q, r) of points from the k-NN list and RNN list of x, compute dist(q, r);
15 | Use $(q, \text{dist}(q, r))$ to update the k-NN list of r, and use $(r, \text{dist}(q, r))$ to update the k-NN list of q;
16 | **end**
17 **until** *maximum number of iterations T is reached*;
18 Return G.

---

4   The LID-Fingerprint Method

LID-Fingerprint is a simple fingerprinting technique that generates a set of compact binary vector representations for the objects of a large, high-dimensional dataset. These binary codes can subsequently be used to select candidates for similarity search, in a filtering phase based on the Hamming distance.

4.1   Object Fingerprinting

As described in Sect. 3.2, For each object x_i in dataset X, we rank the features using the support-weighted ID criterion of Eq. 3, and sparsify the object representation by setting the values of the least locally discriminative features to zero. This process projects each object x_i onto a subset of features $F' \subset F$, with respect to which we denote the new representation by x'_i, where for all $j \in F$, feature $x'_{ij} = x_{ij}$, whenever $j \in F'$, and $x'_{ij} = 0$ otherwise.

The compact representation x'_i of each object $x_i \in X$ can subsequently be transformed into a sparse binary vector x''_i, such that for all $j \in F$, feature $x''_{ij} = 1$, whenever $j \in F'$, and $x'_{ij} = 0$ otherwise. We refer to x''_i as the LID-Fingerprint of x_i.

4.2   Similarity Measure

As a measure of similarity between a query binary code and the fingerprints of dataset objects, we use a restriction of the standard Hamming distance $\mathcal{H}$ to those bits for which the query code is 1. In this restriction, which we call the subspace Hamming distance, the query can be regarded as determining a subset of the features which are deemed to be locally relevant to the object from which the query code has been derived. More formally, the subspace Hamming distance $\mathcal{H}_s$ between a query object q, and any object x_i of X is computed as a bitwise AND operation between the binary code q'' associated with q and the complement $\bar{x}''_i$ of the binary fingerprint of x_i, followed by a bit count on the result:

$$\mathcal{H}_s(x''_q, x''_i) = \sum_{j=1}^{D} (x''_{qj} \ \& \ \bar{x}''_{ij}), \tag{4}$$

where $\&$ denotes the bitwise AND operation. Under the subspace Hamming distance, x_i is considered close to the query q only if x_i has non-zero values over most or all of the features relevant to q, as indicated by the binary code q''.

4.3   Computing Fingerprints

Algorithm 2 illustrates the LID-Fingerprint process. The only changes made to NNWID-Descent (Algorithm 1) involve the recording (as a binary vector for each object) of those features that survive the local sparsification process. In line 2, the binary fingerprints for all objects are initialized to 1. Once the desired proportion of the object features are sparsified, their corresponding fingerprint bits are set to 0 (lines 18–21). Instead of the k-NN graph G (line 22), the algorithm returns the set of binary fingerprints X''.

Algorithm 2: LID-Fingerprint Process (only the modified and the new steps are shown - See Algorithm 1).

Input : Dataset X, distance function dist, neighborhood size k for the graph, neighborhood size K for computing wID scores, sparsification rate Z, number of iterations T.

Output: Binary Fingerprints X''.

2 Normalize the original feature vectors of X;   Initialize X'' to vectors of 1s;

18 **foreach** *data point* $x \in X$ **do**

19 | For each feature f in x, **if** $f \neq 0$ **then** set X''_{x_f} to 1;

20 | **else** set X''_{x_f} to 0

21 **end**

22 Return X''.

4.4 Information Hiding

LID-Fingerprint supports two forms of information hiding: *data suppression*, and *data masking*. Fingerprinting in general can protect the identities, privacy, and personal information by not releasing the actual values of some of the object features, so as to prevent other users from inferring sensitive information. While some of the dataset information is entirely removed in data suppression, in data masking the selected information is concealed or encrypted. The masked data remains encoded in the database, and can be accessed or re-identified only by authorized users.

The replacement of original feature vectors by sparsified binary vectors entails an information loss that helps to preserve the privacy of data. LID-Fingerprint implements data suppression by removing many noisy features locally from each object through the NNWID-Descent feature ranking and sparsification processes. By converting the remaining important features to discrete (binary) values, LID-Fingerprint implements data masking. However, LID-Fingerprint does not guarantee the uniqueness of the generated fingerprints, since any objects sparsified to the same relevant feature subsets would be assigned identical codes. This provides a high degree of anonymity within any query result on the fingerprints, which is likely to defeat any attacker without access to the original feature vectors.

5 Experiments

To test the performance of LID-Fingerprint, we first conducted experiments to study the effectiveness of using fingerprints in place of the original feature vectors, in the accuracy of similarity graph construction. We then tested the indexing performance of LID-Fingerprint against a competing fingerprint generation method based on the SuperBit variant of LSH.

5.1   Datasets

Five real datasets of varying sizes and dimensions were considered:

- **ALOI-100** is a subset of the Amsterdam Library of Object Images (ALOI) [13], which contains 110,250 images of 1000 small objects. Each image is described by a 641-dimensional feature vector based on color and texture histograms. ALOI-100 contains 10,800 images of 100 simple objects generated by selecting the objects uniformly from among the 1000 classes.
- **MNIST** [25] contains 70,000 images, each showing single handwritten numeral (a digit in the range 0 to 9). Each image is represented by 784 gray-scale texture values.
- The **RLCT** (Relative Location of CT) dataset [26] contains 53,500 axial CT slice images from 97 different patients. Each CT slice is described by two histograms in polar space. The feature vectors of the images are of 385 dimensions.
- The **ONP** (Online News Popularity) dataset [12] contains data extracted from 39,644 news articles. Each record has 60 attributes (58 predictive attributes, and 2 non-predictive) that describe different aspects of the source article. The articles are classified as popular or unpopular using a decision threshold over 1400 social interactions.
- The **MiniBooNE** [26] physical particle identification dataset contains 130,065 records of signal and background events. Each event has 50 particle ID variables. This dataset is used to distinguish between two classes, electron neutrinos (signal), and muon neutrinos (background).

5.2   LID Fingerprint Codes and Similarity Search

For each of the five datasets considered in this experiment, we investigated the accuracy of similarity graph construction when using LID-Fingerprint codes as a proxy for the original feature vectors, for the similarity measure defined in Sect. 4.2. The fingerprints were computed over a range of sparsification rates. As a contrast, the accuracy of the similarity graph construction was compared with that of NNWID-Descent for the same sparsification rates.

Parameter Settings. For the contrast experiment between NNWID-Descent and LID-Fingerprint, the similarity graph neighborhood set size was set to $k = 10$. In addition, to show how LID-Fingerprint accuracy scales with k, graphs were also computed for $k = 50$ and $k = 100$. For both NNWID-Descent and LID-Fingerprint, the total proportion of sparsified features per iteration was varied in different ways for different datasets, since the appropriate amount of sparsification depends heavily on the density of the original feature vectors. In each iteration, we set Z to be equal to 0.02 for ONP, 0.04 for MiniBooNE, and 0.0025 for ALOI-100 and the sparse datasets MNIST and RLCT. The nearest neighbor size used for computing the support-weighted ID score, wID, was set to $K = 100$ for all datasets.

Performance Measure. For each of the five datasets, the ground truth class labels of the data objects were used to measure the quality of the resulting k-NN graph at every iteration. The accuracy of the resulting k-NN graph is evaluated, as in [21], using the following formula:

$$graph\ accuracy = (\#correct\ neighbors)/(\#data \times k), \tag{5}$$

where the neighbors deemed to be correct are those that share the same label as the query object. It should be noted that the comparison 'unfairly' favors NNWID-Descent, in that it has access to the original feature values for the unsparsified features, whereas LID-Fingerprint is restricted to binary feature values.

Results and Analysis. For both NNWID-Descent and LID-Fingerprint, Fig. 1 shows the k-NN graph accuracy achieved at every iteration of the methods. For the dense datasets, the plots indicate that when the number of sparsified features is low, the accuracy of the LID-Fingerprint graph is also low; however, this is to be expected, since the binary codes would consist of 1s in all but a handful of locations. As the sparsification rates rise, and the fingerprint codes become more diverse, the accuracy quickly rises. The experiments show that with sufficient sparsification, the performance of LID-Fingerprint codes approaches (and in some cases meets) that of the local feature vectors produced by NNWID-Descent. Although some degradation in performance is to be expected from the use of binary codes over original feature values (whether sparsified or not), the results show that a useful level of accuracy can be achieved. For those applications where both processing speed and anonymity are essential, indexing based on LID-Fingerprint codes can lead to good performance, particularly when followed by refinement of the query result.

5.3   LID-Fingerprint Versus SuperBit LSH

In this experiment, we tested the performance of LID-Fingerprint codes for k-NN search, in terms of both accuracy and execution time, against the SuperBit variant of LSH [22]. SuperBit improves the LSH random projection strategy by computing an estimation of cosine similarity, and orthogonalizing the random projection vectors in batches. SuperBit computes a binary hash code for each input data vector. For the experiment, we used the Debatty v0.11 Java implementation of SuperBit available at GitHub.

For each method and each of the datasets considered, we randomly selected 1000 objects to serve as queries. For LID-Fingerprint, k-NN search was implemented by precomputing fingerprint codes using Algorithm 2, and then performing a sequential search of the full fingerprint dataset at query time. For SuperBit, k-NN search was implemented using standard LSH bucket searching techniques.

Parameter Settings. For LID-Fingerprint preprocessing, fingerprint code sets were generated using the same parameter settings specified in Sect. 5.2, with the

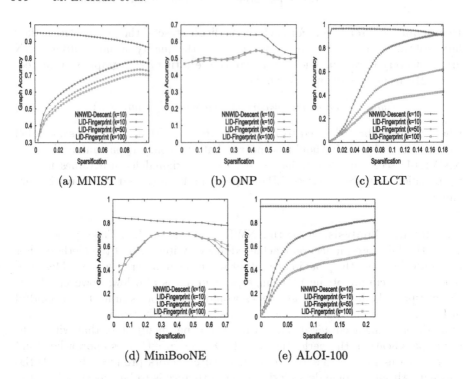

Fig. 1. Comparison between NNWID-Descent similarity graph ($k = 10$) and LID-Fingerprint similarity graph for different values of k (10, 50, and 100). The y-axis shows average neighborhood accuracy with respect to ground truth classes, while the x-axis indicates the proportion of features sparsified.

exception that for the ONP and MiniBooNE datasets, the number of iterations T was set to 14 and 12, respectively (since the highest similarity graph accuracies in Fig. 1 were obtained for these settings).

For the SuperBit method, the number of hash functions was set to 2 for those datasets having small dimension D (ONP and MiniBooNE), and to 10 for those datasets of larger dimension (MNIST, RLCT, and ALOI-100). For all datasets, the number of buckets was set equal to the number of ground truth classes.

Performance Measure. Two evaluation parameters were measured as an average over all queries: accuracy, and search time. The search time was calculated as the proportion of the time required to compute a query k-NN result (excluding preprocessing). For an individual query q, the accuracy of its k-NN result is defined as:

$$query\ accuracy = (\#correct\ neighbors)/k, \qquad (6)$$

where the neighbors deemed to be correct are those that share the same label as the query object.

Results and Analysis. Figure 2 shows the average accuracy and time achieved for 10-NN search of LID-Fingerprint and SuperBit LSH. For all datasets considered, LID-Fingerprint sequential search outperforms SuperBit LSH in terms of both accuracy and execution time. SuperBit accuracy could be increased by increasing the number of buckets, but only at the expense of execution time; conversely, the execution time could be improved by reducing the number of buckets at the expense of accuracy. For this reason, for the datasets studied, we conclude that the performance of similarity search with LID-Fingerprint codes is superior to that of SuperBit LSH, even when limiting the LID-Fingerprint search strategy to sequential search.

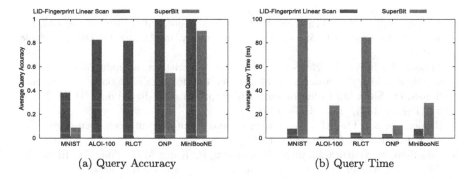

(a) Query Accuracy   (b) Query Time

Fig. 2. The 10-NN search performance of LID-Fingerprint (sequential search) versus SuperBit LSH (bucket search), in terms of the average accuracy and time over 1000 queries.

6   Conclusion and Future Work

In this paper, we presented LID-Fingerprint, a simple binary fingerprinting technique derived as a byproduct of the use of a local intrinsic dimensionality criterion, Support-Weighted Intrinsic Dimensionality (support-weighted ID), to guide the NNWID-Descent similarity graph construction method. Using several high-dimensional real datasets, experimental results have shown that LID-Fingerprint can be applied to generate binary codes that provide a reasonable level of search accuracy when compared with that of the original feature representations. In comparison with state-of-the-art method for fingerprint generation (SuperBit LSH), LID-Fingerprint supports similarity search with very competitive performance in both accuracy and execution time.

LID-Fingerprint has the potential to provide secure search as well as reduce the storage and computational complexity for many cloud and smart-phone applications. In addition to providing data suppression and data masking, LID-Fingerprint also provides a reasonable level of data anonymity. However, with large datasets, the number of generated fingerprints can be high, and similarity queries can be expensive to compute. Thus, accelerating search operations through compression, as well as developing robust and scalable indexing and refinement methods for sparse fingerprint codes, are interesting directions for future research.

Acknowledgments. M. E. Houle acknowledges the financial support of JSPS Kakenhi Kiban (B) Research Grant 18H03296, and V. Oria acknowledges the financial support of NSF Research Grants DGE 1565478 and AGS 1743321.

References

1. Amsaleg, L., Chelly, O., Furon, T., Girard, S., Houle, M.E., Kawarabayashi, K., Nett, M.: Estimating local intrinsic dimensionality. In: ACM SIGKDD, pp. 29–38 (2015)
2. Andoni, A., Indyk, P.: Near-optimal hashing algorithms for approximate nearest neighbor in high dimensions. In: FOCS, pp. 459–468. IEEE (2006)
3. Boneh, D., Shaw, J.: Collusion-secure fingerprinting for digital data. IEEE Trans. Inf. Theory **44**(5), 1897–1905 (1998)
4. Broder, A.Z.: On the resemblance and containment of documents. In: SEQUENCES, pp. 21–29. IEEE (1997)
5. Caballero, J., Venkataraman, S., Poosankam, P., Kang, M.G., Song, D., Blum, A.: FiG: automatic fingerprint generation. In: NDSS Symposium (2007)
6. Cano, P., Batlle, E., Kalker, T., Haitsma, J.: A review of audio fingerprinting. J. VLSI Sig. Process. Syst. **41**(3), 271–284 (2005)
7. Charikar, M.S.: Similarity estimation techniques from rounding algorithms. In: ACM STOC, pp. 380–388 (2002)
8. Cox, I., Miller, M., Bloom, J., Fridrich, J., Kalker, T.: Digital Watermarking and Steganography, 2nd edn. Morgan Kaufmann Publishers Inc., San Francisco (2008)
9. Dong, W., Moses, C., Li, K.: Efficient K-nearest neighbor graph construction for generic similarity measures. In: WWW, pp. 577–586 (2011)
10. Farooq, F., Bolle, R.M., Jea, T.Y., Ratha, N.: Anonymous and revocable fingerprint recognition. In: IEEE CVPR, pp. 1–7 (2007)
11. Fei-Fei, L., Perona, P.: A bayesian hierarchical model for learning natural scene categories. In: IEEE CVPR, vol. 2, pp. 524–531, June 2005
12. Fernandes, K., Vinagre, P., Cortez, P.: A proactive intelligent decision support system for predicting the popularity of online news. In: Pereira, F., Machado, P., Costa, E., Cardoso, A. (eds.) EPIA 2015. LNCS (LNAI), vol. 9273, pp. 535–546. Springer, Cham (2015). https://doi.org/10.1007/978-3-319-23485-4_53
13. Geusebroek, J.M., Burghouts, G.J., Smeulders, A.W.M.: The Amsterdam library of object images. IJCV **61**(1), 103–112 (2005)
14. Gionis, A., Indyk, P., Motwani, R.: Similarity search in high dimensions via hashing. VLDB **99**, 518–529 (1999)
15. Gong, Y., Lazebnik, S.: Iterative quantization: a procrustean approach to learning binary codes. In: IEEE CVPR, pp. 817–824 (2011)
16. Halpern, J.Y., O'Neill, K.R.: Anonymity and information hiding in multiagent systems. J. Comput. Secur. **13**(3), 483–514 (2005)

17. Houle, M.E.: Local intrinsic dimensionality I: an extreme-value-theoretic foundation for similarity applications. In: Beecks, C., Borutta, F., Kröger, P., Seidl, T. (eds.) SISAP 2017. LNCS, vol. 10609, pp. 64–79. Springer, Cham (2017). https:// doi.org/10.1007/978-3-319-68474-1_5

18. Houle, M.E., Oria, V., Wali, A.M.: Improving k-NN graph accuracy using local intrinsic dimensionality. In: Beecks, C., Borutta, F., Kröger, P., Seidl, T. (eds.) SISAP 2017. LNCS, vol. 10609. Springer, Cham (2017). https://doi.org/10.1007/ 978-3-319-68474-1_8

19. Houle, M.E.: Dimensionality, discriminability, density and distance distributions. In: IEEE ICDMW, pp. 468–473 (2013)

20. Houle, M.E.: Local intrinsic dimensionality II: multivariate analysis and distributional support. In: Beecks, C., Borutta, F., Kröger, P., Seidl, T. (eds.) SISAP 201. LNCS, vol. 10609. Springer, Cham (2017). https://doi.org/10.1007/978-3-319-68474-1_6

21. Houle, M.E., Ma, X., Oria, V., Sun, J.: Improving the quality of K-NN graphs through vector sparsification: application to image databases. IJMIR 3(4), 259–274 (2014)

22. Ji, J., Li, J., Yan, S., Zhang, B., Tian, Q.: Super-bit locality-sensitive hashing. In: NIPS, pp. 108–116 (2012)

23. Katzenbeisser, S., Petitcolas, F.: Information Hiding. Artech House (2016)

24. Lampson, B.W.: A note on the confinement problem. Commun. ACM 16(10), 613–615 (1973)

25. Lecun, Y., Bottou, L., Bengio, Y., Haffner, P.: Gradient-based learning applied to document recognition. Proc. IEEE 86(11), 2278–2324 (1998)

26. Lichman, M.: UCI Machine Learning Repository (2013). http://archive.ics.uci. edu/ml

27. Mason, L., Baxter, J., Bartlett, P.L., Frean, M.R.: Boosting algorithms as gradient descent. In: NIPS, pp. 512–518 (2000)

28. Moravec, K., Cox, I.J.: A comparison of extended fingerprint hashing and locality sensitive hashing for binary audio fingerprints. In: ACM ICMR, p. 31 (2011)

29. Petitcolas, F.A., Anderson, R.J., Kuhn, M.G.: Information hiding-a survey. Proc. IEEE 87(7), 1062–1078 (1999)

30. Raginsky, M., Lazebnik, S.: Locality-sensitive binary codes from shift-invariant kernels. In: NIPS, pp. 1509–1517 (2009)

31. Salakhutdinov, R., Hinton, G.: Semantic hashing. RBM 500(3), 500 (2007)

32. Shakhnarovich, G., Viola, P., Darrell, T.: Fast pose estimation with parameter-sensitive hashing. In: IEEE ICCV, p. 750 (2003)

33. Strecha, C., Bronstein, A., Bronstein, M., Fua, P.: LDAhash: improved matching with smaller descriptors. TPAMI 34(1), 66–78 (2012)

34. Torralba, A., Fergus, R., Weiss, Y.: Small codes and large image databases for recognition (2008)

35. Weiss, Y., Torralba, A., Fergus, R.: Spectral hashing. In: NIPS, pp. 1753–1760 (2009)

36. Xu, H., Veldhuis, R.N.: Binary representations of fingerprint spectral minutiae features. In: IEEE ICPR, pp. 1212–1216 (2010)

37. Xu, H., Veldhuis, R.N., Kevenaar, T.A., Akkermans, T.A.: A fast minutiae-based fingerprint recognition system. IEEE Syst. J. 3(4), 418–427 (2009)

Clustering and Outlier Detection

Applying Compression to Hierarchical Clustering

Gilad Baruch[1], Shmuel Tomi Klein[1(✉)], and Dana Shapira[2]

[1] Department of Computer Science, Bar Ilan University, 5290002 Ramat Gan, Israel
gilad.baruch@gmail.com, tomi@cs.biu.ac.il
[2] Department of Computer Science, Ariel University, 40700 Ariel, Israel
shapird@ariel.ac.il

Abstract. Hierarchical Clustering is widely used in Machine Learning and Data Mining. It stores bit-vectors in the nodes of a k-ary tree, usually without trying to compress them. We suggest a data compression application of hierarchical clustering with a double usage of the xoring operations defining the Hamming distance used in the clustering process, extending it also to be used to transform the vector in one node into a more compressible form, as a function of the vector in the parent node. Compression is then achieved by run-length encoding, followed by optional Huffman coding, and we show how the compressed file may be processed directly, without decompression.

1 Introduction

A common problem appearing in various different application fields, may be generically described as follows. We assume the existence of a large collection S of elements which is stored on our computer, possibly after having been pre-processed to facilitate access and searches. Given a query q, which is an element that may, or may not, belong to S, the problem is to locate an element of S which is most similar or *closest* to q. If closeness is defined as identity, this is just a standard search problem for an element in a set. In a more general setting, the *distance* between elements could be taken into account, for some appropriate definition of a metric. This has many applications, such as Document Retrieval systems, finding the most similar picture within an image database, locating similar reads in bioinformatics applications, as well as various problems in Machine Learning, to mention just a few.

An exact solution to this *Nearest Neighbor* problem is hard to obtain [20], but many approximate schemes have been suggested. One of these approaches is to partition the set S into disjoint parts C_i called *clusters*, with $S = C_1 \cup C_2 \cup \cdots$, so that pairs of elements within the same cluster C_i have a higher similarity than pairs of elements belonging to different clusters. The idea is then, given a query q, to first locate the most fitting cluster C_i and to search for q within C_i in a second stage.

Applying the clustering idea recursively to each of the clusters themselves yields a hierarchical structure known as *Hierarchical Clustering* [9,17] which is

© Springer Nature Switzerland AG 2018
S. Marchand-Maillet et al. (Eds.): SISAP 2018, LNCS 11223, pp. 151–162, 2018.
https://doi.org/10.1007/978-3-030-02224-2_12

commonly used in Machine Learning and Data Mining [18,20], and is considered as one of the best approximate Nearest Neighbor schemes [15]. It is a k-ary tree, in which each leaf corresponds to one of the elements of the set S and stores a bit-vector of length b associated with some feature vector defining that element. We shall identify these bit-vectors with the corresponding elements of the set S they represent, so that the query q is also represented as a bit-vector.

The internal nodes of the tree also store bit-vectors, specifically, the vector stored in an internal node v will be a representative of the subset of S whose bit-vectors are stored in the leaves of the subtree rooted at v. The standard way of defining a representative bit-vector is to apply the majority rule to each bit position of the constituent bit-vectors.

In the field of robotics and smart devices, where real-time performance under strict budget constraints is crucial, the traditional feature-descriptors used for Computer Vision tasks such as SIFT (Scale Invariant Feature Transform) [14] and SURF (Speeded Up Robust Features) [2] are being replaced by the real-time family of the simpler *binary* descriptors (such as BRIEF (Binary Robust Independent Elementary Features) [5], ORB (Oriented and Rotated BRIEF) [19] and BRISK (Binary Robust Invariant Scalable Keypoints) [13], to name just a few). Binary descriptors in general are representing a patch in an image by comparing the intensities of pairs of pixels in the patch. In the case that the value at the first pixel of the pair is larger than at the other pixel, a 1-bit is assigned, and a 0-bit otherwise. The results of the comparisons between many different pairs (usually 256 or 512) in the patch are being concatenated to form a bit-vector, which is the binary descriptor.

We focus on Hierarchical clustering constructed according to the *Hamming distance*. The Hamming distance of two bit-vectors X and Y of length n is the number of bit substitutions required in order to transform one bit-vector into the other. This operation can be efficiently computed by

$$HD(X, Y) = \text{number of 1-bits in the bit-vector}(X \text{ xor } Y).$$

An example of a Hierarchical clustering with $k = 3$ is given in Fig. 1(a). Many application such as SLAM (Simultaneous Localization and Mapping) [16] and Place Recognition [10] are using binary descriptors in a structure of a hierarchical clustering tree. Those applications require using the minimal amount of space possible, while still allowing real-time operations to be performed.

Typical Hierarchical Clustering constructions use the bit-vectors in their original form, despite the fact that these files tend to be quite large. The omission of compression is justified by the high entropy of these data files for the binary alphabet. Nevertheless, in several Computer Vision applications, the memory bandwidth is limited. Consequently, reducing the size of the file could result in a faster initialization, as the file is copied in less time to main memory.

Moreover, given the memory limitations, large files cannot be stored on the device itself, and have therefore to be imported. We suggest here applying a compression component to the Hierarchical Clustering structure, and show how searches can be performed directly on the compressed form. Our experiments show that our new method not only saves space, but also offers superior processing times. Similar research was carried out in [12], where compression was applied to sets of feature vectors generated by SIFT, see [14], under the constraint of processing the data directly in its compressed form. Two dimensional compressed pattern matching was addressed in [11] for searching JPEG encoded images.

A similar tree structure to the one we suggest is presented in [7], though with different objectives and for different applications. They consider ordered lists of integers, and use the tree to facilitate access and search, while we deal with vectors and search for nearest neighbors.

The paper is organized as follows. Our method is presented in three stages: Sect. 2 recalls the basic hierarchical clustering and suggests a different representation of the hierarchical structure so that the data is correlated and therefore compressible, without harming the searching process. This is reminiscent of the Burrows-Wheeler Transform [4], which by itself is not a compression method and only recodes the input into a more compressible file. Similarly, the outcome of the first stage of our algorithm is sparser, suggesting an ad-hoc *Run Length Encoding* (RLE), which is given in Sect. 3. To gain additional compression, the sequences generated by the RLE may be Huffman encoded, though at the price of losing the ability of directly searching the compressed file. Section 4 presents experimental results.

2   Hierarchical Clustering Compression

The search for a query item q, represented by its bit-vector B_q, starts at the root of the hierarchical clustering tree. Navigation proceeds to the most similar child of the current node, until a leaf is reached. The most similar child is defined as that whose bit-vector has a minimal distance to B_q, among the vectors stored in the children of the current node.

The formal search algorithm for querying a hierarchical structure T rooted at $root(T)$ is presented in Algorithm 1. As example, let $B_q = 1011\ 1101\ 0110$, and refer to the Hierarchical clustering presented in Fig. 1(a). The search starts at the root of T and examines the distances between all of its children and B_q, which are, from left to right, 2, 4 and 7. The procedure thus recurses on the left child of the root, computes the distances of its children to B_q as 4, 5 and 2, and thus returns its rightmost child, having HD = 2 from B_q.

At a first look, the entropy of the data resulting from the hierarchical clustering process is close to that of random data, meaning that no significant compression can be achieved, since truly random data is known to be incompressible. However, the hierarchical structure is achieved by clustering similar bit-vectors in a parent and child relation. Exploiting this similarity, we can assume that many bits of a child's data are identical to that of its parent node. Following ideas from [3], we present in this section the hierarchical structure using a method based on xoring (applying eXclusive OR) the vector of node v with that of v's parent node. Note that we make double use of the xor operator, for two seemingly independent purposes: it serves both as a compression method for the involved data structure, as well as a component of the Hamming distance we wish to minimize while processing a query.

2.1   A Sparser Hierarchical Clustering Tree

For every node v which is not the root, instead of storing the bit-vector B_v of v itself, we propose to store the result $\hat{B}_v$ of xoring B_v with the bit-vector $B_{p(v)}$ of the parent $p(v)$ of v in the hierarchical tree, that is,

$$\hat{B}_v = B_v \text{ xor } B_{p(v)}. \tag{1}$$

If the vectors are indeed similar, the resulting vector $\hat{B}_v$ may be much sparser than the original B_v. Therefore, general compression techniques will typically perform better on the file constructed according to this strategy. The original data B_v may be restored by using the same xor operation once again, i.e., by applying

$$\hat{B}_v \text{ xor } B_{p(v)} = \big((B_v \text{ xor } B_{p(v)}) \text{ xor } B_{p(v)}\big) = B_v.$$

For example, let $B_v = 1011\ 1101\ 0101$ and $B_{p(v)} = 1010\ 1101\ 1101$ be two correlated bit-vectors. The resulting stored bit-vector is $\hat{B}_v = B_v \text{ xor } B_{p(v)} = 0001\ 0000\ 1000$, which is sparser than the vector B_v of the child. Figure 1 presents an example of Hierarchical clustering, before and after the xoring process has been applied. The root of the tree in Fig. 1(b) is the same as for Fig. 1(a), the vectors in the other nodes are obtained from their counterparts by xoring with the vectors in their respective parent nodes.

If several layers of xoring are applied, recovering the original data is not immediate, but in fact, it is not really needed. The structure is used to locate an element of the set $\mathcal{S}$ which is closest to a given query, and we shall show how this element can be located working only with the xored bit-vectors and circumventing their original, non-xored counterparts.

The modified algorithm for querying T is presented in Algorithm 2. The difference with Algorithm 1 other than using sparser xored vectors $\hat{B}_v$ instead of the original B_v, is in the addition of lines 2 and 11, in which the query vector $\tilde{B}_q$ is constantly updated.

Returning to our running example with $B_q = 1011\ 1101\ 0110$, we now refer to the XORed Hierarchical clustering presented in Fig. 1(b). Although in Algorithm 2 the Hamming distance is applied on other vectors than in Algorithm 1, the distances remain the same. Searching again starts at the root of T, but this time the distances are computed between the children $\hat{B}_w$ of the root and the modified query vector $\tilde{B}_q = 0001\ 0000\ 1011$, yielding the same values as before, 2, 4 and 7. The same is true for the comparisons on the next level, where $\tilde{B}_q$ gets updated to $0000\ 0000\ 0011$, with distances 4, 5 and 2, finally returning the leaf $0010\ 0000\ 0010$, whose non-xored counterpart $1001\ 1101\ 0111$ has Hamming distance 2 from B_q.

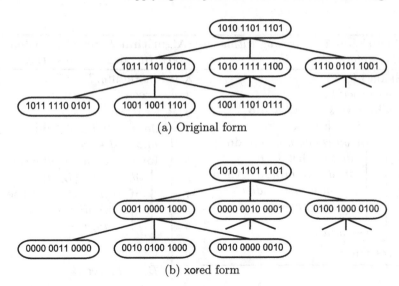

(a) Original form

(b) xored form

Fig. 1. Hierarchical clustering.

2.2 Correctness of the Proposed Search

To understand the correctness of the procedure, consider the following claims.

Lemma 1: The bit-vector $\tilde{B}_q$ used as the querying vector in line 6 of Algorithm 2 is

$$\tilde{B}_q = B_q \text{ xor } B_v, \tag{2}$$

which is the xor of the original vector of the query q and the original (not xored) vector stored in the node v.

Proof: Denote the sequence of nodes assigned to the variable v during the execution of the algorithm by $V = \{v_1, v_2, \ldots, v_r\}$, where $v_1 = root(T)$ and v_r is a leaf. The proof of the lemma is by induction on the depth of the currently visited node in the tree, which is the index i of the current element v_i in the sequence V. For $i = 1$, Eq. (2) holds because of the initializations in lines 1 and 2.

Suppose the claim holds after i iterations. The value of $\tilde{B}_q$ after the ith application of line 6 is thus B_q xor B_{v_i}. In line 10, the value of v_{i+1} is then set as the child of v_i for which the minimum is attained, and in line 11, $\tilde{B}_q$ is updated to be

$$\tilde{B}_q \leftarrow \tilde{B}_q \text{ xor } \hat{B}_{v_{i+1}} = (B_q \text{ xor } B_{v_i}) \text{ xor } (B_{v_{i+1}} \text{ xor } B_{v_i}) = B_q \text{ xor } B_{v_{i+1}},$$

where the first parentheses are due to the inductive assumption, and the second to the definition of $\hat{B}_{v_{i+1}}$ as given in Eq. (1). The new value assigned to $\tilde{B}_q$ in line 11 of iteration i is its value in line 6 of iteration $i + 1$, which shows that the claim also holds after $i + 1$ iterations.   ∎

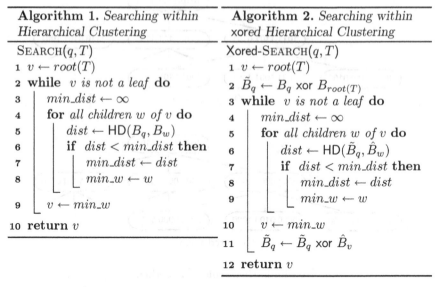

Algorithm 1. *Searching within Hierarchical Clustering*

SEARCH(q, T)
1 $v \leftarrow root(T)$
2 **while** v *is not a leaf* **do**
3     $min\_dist \leftarrow \infty$
4   **for** *all children* w *of* v **do**
5        $dist \leftarrow \mathsf{HD}(B_q, B_w)$
6   **if** $dist < min\_dist$ **then**
7           $min\_dist \leftarrow dist$
8           $min\_w \leftarrow w$
9     $v \leftarrow min\_w$
10 **return** v

Algorithm 2. *Searching within xored Hierarchical Clustering*

Xored-SEARCH(q, T)
1 $v \leftarrow root(T)$
2 $\tilde{B}_q \leftarrow B_q$ xor $B_{root(T)}$
3 **while** v *is not a leaf* **do**
4     $min\_dist \leftarrow \infty$
5   **for** *all children* w *of* v **do**
6        $dist \leftarrow \mathsf{HD}(\tilde{B}_q, \hat{B}_w)$
7   **if** $dist < min\_dist$ **then**
8           $min\_dist \leftarrow dist$
9           $min\_w \leftarrow w$
10    $v \leftarrow min\_w$
11    $\tilde{B}_q \leftarrow \tilde{B}_q$ xor $\hat{B}_v$
12 **return** v

Lemma 2: For all bit-vectors A, B and C, all of the same length,

$$\mathsf{HD}(A, B) = \mathsf{HD}(A \text{ xor } C, B \text{ xor } C).$$

Proof: Immediate from the definition of the Hamming distance and the properties of the bit-wise xor operation, since

$$(A \text{ xor } C) \text{ xor } (B \text{ xor } C) = A \text{ xor } B. \qquad \blacksquare$$

Theorem: Given any clustering tree T and any query q, Algorithms 1 and 2 return the same value v.

Proof: This follows from the fact that the distances $dist$ calculated in line 5 of Algorithm 1 and in line 6 of Algorithm 2 are in fact the same. Indeed, applying the above lemma and the definition of the actually stored bit-vector, one gets

$$\mathsf{HD}(\tilde{B}_q, \hat{B}_w) = \mathsf{HD}(B_q \text{ xor } B_v, B_w \text{ xor } B_v) = \mathsf{HD}(B_q, B_w). \qquad \blacksquare$$

The interesting feature of Algorithm 2 is thus that it processes only xored bit-vectors, which are sparser and therefore may be compressed more efficiently, yet calculates the same Hamming distances as before and thus is able, given a query, to correctly locate its closest element.

3   Run Length Encoding

The purpose of using xored vectors is to get sparser ones that are more compressible. We use *Run Length Encoding* (RLE) on the binary alphabet, i.e., enumerating the lengths of runs of 0-bits, to encode the outcome of the hierarchical clustering process. A *run* of zeros of length r, $r \geq 0$, is a sequence of r zeros followed by a 1-bit, unless it is the last run in a sequence, in which case the terminating 1-bit is omitted. For example, the binary string 0011000001000 is represented by the sequence $(2, 0, 5, 3)$, whereas the string 00011100001 would be encoded as $(3, 0, 0, 4, 0)$, as it needs a final zero to indicate that the string finishes with a 1-bit.

Actually, the last run length need not be encoded in the case that the total length b of the encoded bit-vector is known, since it can be deduced. Indeed, if the RLE is $(l_1, l_2, \ldots, l_s)$, then $(\sum_{i=1}^{s} l_i) + s - 1 = b$ so that one can restore the last element of the RLE sequence as a function of those preceding by

$$l_s = b - s + 1 - \sum_{i=1}^{s-1} l_i.$$

The following algorithm RL-HD shows how to compute the Hamming Distance of two RLE encoded binary strings. It is given two bit-vectors, represented by the vectors holding their RLE encodings $l[1 : s]$ and $d[1 : r]$, where we assume that their last elements have already been restored. Note that the running time of RL-HD is proportional to the size of the run length encoding, and not the decompressed size, that is not proportional to the total number of bits, but only to the number of 1-bits. There is thus an advantage when the bit-vectors are sparse, as we assume here.

Algorithm 3 could be used as part of the searching algorithm presented in Algorithm 1. Instead of the assignment to *dist* in line 5 of Algorithm 1, the command

$$dist \leftarrow \mathsf{RL-HD}\ (l[1 : s], d[1 : r])$$

is applied with $l = \mathsf{RLE}\ (B_q)$, $d = \mathsf{RLE}\ (B_w)$, where $\mathsf{RLE}(X)$ denotes the RLE encoding of the bit-vector X.

A similar adaptation could be done to Algorithm 2 which works with the xored vectors, but there is a problem with $\tilde{B}_q$ which changes dynamically in each iteration, whereas B_q in Algorithm 1 was constant. To be able to apply the xored search of Algorithm 2 directly with the RLE encodings, not only $\tilde{B}_q$ has to be updated, but also its RLE encoding. This is done by Algorithm 4, which gets the RLE encodings of two bit-vectors X and Y as input in the same format as RL-HD, and returns the RLE encoding of X XOR Y as an array $z[1 : k]$. The update of $z[k]$ in the last line before the return is because the final run does not terminate in a 1-bit. Lines 2, 6 and 11 of Algorithm 2 have thus to be updated to

2      $\mathsf{RLE}(\tilde{B}_q) \leftarrow \mathsf{RL-xor}\ \big(\mathsf{RLE}(B_q), \mathsf{RLE}(B_{root(T)})\big)$

6           $dist \leftarrow \mathsf{RL-HD}\ \big(\mathsf{RLE}(\tilde{B}_q), \mathsf{RLE}(\hat{B}_w)\big)$

11     $\mathsf{RLE}(\tilde{B}_q) \leftarrow \mathsf{RL-xor}\ \big(\mathsf{RLE}(\tilde{B}_q), \mathsf{RLE}(\hat{B}_v)\big)$

Algorithm 3. *Hamming Distance on RLE encoded vectors*	**Algorithm 4.** *RLE of the* xor *of two bit-vectors*
RL-HD($l[1:s], d[1:r]$)	RL-xor($l[1:s], d[1:r]$)
1 $dist \leftarrow 0$	1 $z[1] \leftarrow 0$
2 $i \leftarrow 1 \qquad j \leftarrow 1$	2 $i \leftarrow 1 \qquad j \leftarrow 1 \qquad k \leftarrow 1$
3 **while** $i \leq s$ **and** $j \leq r$ **do**	3 **while** $i \leq s$ **and** $j \leq r$ **do**
4 $\quad$ **if** $l[i] = d[j]$ **then**	4 $\quad$ **if** $l[i] = d[j]$ **then**
5 $\qquad$ i++ $\qquad j$++	5 $\qquad$ $z[k] \leftarrow z[k] + l[i] + 1$
6 $\quad$ **else**	6 $\qquad$ i++ $\qquad j$++
7 $\qquad$ **if** $l[i] > d[j]$ **then**	7 $\quad$ **else**
8 $\qquad\quad$ $l[i] \leftarrow l[i] - (d[j] + 1)$	8 $\qquad$ **if** $l[i] > d[j]$ **then**
9 $\qquad\quad$ j++	9 $\qquad\quad$ $z[k] \leftarrow z[k] + d[j]$
10	10 $\qquad\quad$ $l[i] \leftarrow l[i] - (d[j] + 1)$
11 $\qquad$ **else**	11 $\qquad\quad$ j++
12 $\qquad\quad$ $d[j] \leftarrow d[j] - (l[i] + 1)$	12 $\qquad$ **else**
13 $\qquad\quad$ i++	13 $\qquad\quad$ $z[k] \leftarrow z[k] + l[i]$
14	14 $\qquad\quad$ $d[j] \leftarrow d[j] - (l[i] + 1)$
15 $\qquad$ $dist$++	15 $\qquad\quad$ i++
16 **return** $dist$	16 $\qquad$ k++
	17 $z[k] \leftarrow z[k] - 1$
	18 **return** $z[1:k]$

4 Experimental Results

Recall that the number of children of each internal node is constant and equal to k. Let n denote the number of leaves in the hierarchical structure, which is the number of samples we can search for. If we assume for simplicity that the hierarchy is a full k-ary tree, then the total number of HD evaluations during a single search is equal to $k \log_k n$, as the distance should be computed to all k children of every node on the path going from the root to the requested leaf. Having a larger k increases the number of distance operations applied within each layer, but decreases the tree's height, $\log_k n$. To compute the best possible k for this model, we fix n and define $f(k) = k \log_k n$, which is minimized at $k = e$. We therefore used the branching parameters $k = 2$ and $k = 3$, but also checked the traditional values $k = 10$ and $k = 16$ in our experiments, which gave inferior results.

In a first set of experiments, we applied the hierarchical clustering to real life data in order to evaluate the compression gains. We used a known image database called COCO – Common Objects in Context[1], from which the first 10000 images of the 2014 Train images set have been extracted. These were fed into a Bag-of-word image database for image retrieval known as the DBoW2 library [10], which

---

[1] http://cocodataset.org/#download.

generated 500 descriptors per image using the ORB feature extraction algorithm
[19]. Each descriptor is a bit-vector of length either 256 or 512.

The DBoW2 software then performs hierarchical clustering by applying k-means++ [1], and finally cleans the data by purging duplicates and replacing sub-trees exceeding some predetermined depth by the centroid of their leaves. All experiments were conducted on a machine running 64 bit Linux Ubuntu with an Intel Core i5-7440 at 2.8 GHz processor, 6144K L3 cache size of the CPU, and 8GB of main memory. The results are presented in Table 1. The first two columns show the sizes of the feature vectors (256 or 512) and the values of k we tested (2 or 3). The next columns give the number of vectors in each dataset, the density of 1-bits in the xored bit-vectors and the total size of the files in MB. The last two columns show the compression performance, as a ratio relative to the uncompressed size. The penultimate column is the result of using fixed length encoded run lengths (using the maximal needed 8 bits for 256 and 9 bits for 512). Just about 30% have been saved for $k = 2$, and even less, 8%, for $k = 3$, because the involved bit-vectors are not sparse enough, with a density of 8–11%. However, the distribution of the run-lengths is strongly biased toward the shorter ones, and applying Huffman coding then yields 55–66% savings.

Table 1. Compression results on test files.

Size in bit of feature vector	k	Number of vectors in million	xored density of 1-bits in %	Total file size in MB	Relative size of fixed RLE	Relative size of Huff on RLE
256	2	1.25	8.6	38.29	0.687	0.358
256	3	1.44	11.5	44.07	0.919	0.457
512	2	1.02	7.8	62.55	0.704	0.335
512	3	1.19	10.2	72.80	0.920	0.424

We also tried other compression techniques for sparse bit-vectors, from Huffman coding on blocks of bits, combined with a special encoding for runs of zero-blocks [8], to various variants of hierarchical compression [6]. The results for the different methods were quite close, but on the given dataset, the simple Huffman coding on run-lengths gave consistently the best results.

The purpose of the following set of tests was to empirically compare the RLE based Hamming distance of Algorithm 3 with a standard evaluation of the HD, based on xoring the arguments and counting the number of 1-bits in the resulting vector, an operation known as popcount on some compilers. The results obviously depend on the density of the 1-bits, which has been set as main parameter, varying its value from $\frac{1}{4}$ to 2^{-9} and halving at each step. For each density p, 1000 bit-vectors $B_1, \ldots, B_{1000}$ of length 512 bits have been generated randomly with probability p for the appearance of a 1-bit, all bit generations being independent from each other. Then the Hamming distance $HD(B_i, B_j)$ has

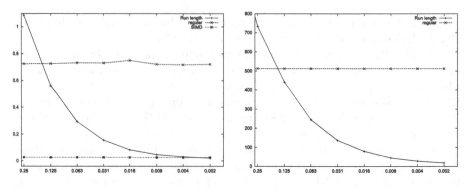

Fig. 2. Average time vs. 1-bit density.   **Fig. 3.** Average size vs. 1-bit density.

been calculated for each pair $1 \leq i, j \leq 1000$ and the time of these evaluations has been summed. This was repeated for each of the considered values of p.

Figure 2 plots the time in microseconds, averaged per bit-vector pair, as a function of the 1-bit density. We see that the time for the regular HD evaluation is almost constant, whereas the RLE based one decreases with decreasing density and performs better for densities of 12% and below. The regular evaluation though can be accelerated by the use of SIMD instructions, which process several bytes as a bulk, if they are supported. As can be seen, the RLE based HD outperforms even SIMD implementations for densities as low as 0.2%.

Figure 3 shows the average size of the vectors in bits as a function of the 1-bit density. The straight line at 512 corresponds to the uncompressed vectors used by the regular HD evaluation. Fixed length encoding, with $9 = \log_2(512)$ bits for each run-length, has been used. Obviously, one could have used Huffman coding to improve the space savings, but that would then require an additional decoding phase which impairs the processing time.

Table 2. Average timing results in microseconds on test files.

Size	k	Alg 1	Alg 2	HD-RLE	Hybrid	Huffman
256	2	11.38	12.77	17.32	13.90	40.55
256	3	11.32	11.58	19.29	14.71	46.19
512	2	20.89	21.86	27.86	24.95	62.07
512	3	21.37	27.13	32.45	25.01	74.42

To get the timing results of Table 2, the same files as for Table 1 have been used, 10000 randomly generated query vectors were used and the running times averaged. As expected, working on the xored vectors in Algorithm 2 took slightly more time than processing the original, uncompressed vectors, because of the repeated update of the query bit-vector $\tilde{B}_q$. In our case, the direct evaluation of the HD given the RLE took 30–70% more time than Algorithm 1, and if Huffman coding is applied, the time was up to 3–4 times worse. Motivated by the compression gain of the xored hierarchical structure at the cost of an increase in search time, relative to processing the decompressed data, we suggest a compromise. In a *hybrid method*, the original bit-vectors are used for the first few levels, and the RLE of the xored ones is used in the lower levels, because the 1-bit density for the upper levels of the tree is very high, around 50% for the upper 4 levels, but drops to about 8% below. Obviously, there is no loss in compression efficiency, as the high density of the upper levels may even trigger negative compression when the RLE encoding of a bit-vector is larger than the original one. The savings are however negligible, as only a small portion of the nodes is involved. The performance of the hybrid method is only 20–30% slower than that of Algorithm 1, which is applied on the uncompressed data, and does not take into account the time spent to decompress the data involved. We conclude that the advantage of the suggested method, for our test files, is space savings, which could imply, for applications with lower 1-bit density, also better processing times.

References

1. Arthur, D., Vassilvitskii, S.: *k*-means++: the advantages of careful seeding. In Proceedings of the Eighteenth Annual ACM-SIAM Symposium on Discrete Algorithms, SODA 2007, New Orleans, Louisiana, USA, January 7–9, 2007, pp. 1027–1035, 2007
2. Bay, H., Tuytelaars, T., Van Gool, L.: SURF: speeded up robust features. In: Leonardis, A., Bischof, H., Pinz, A. (eds.) ECCV 2006. LNCS, vol. 3951, pp. 404–417. Springer, Heidelberg (2006). https://doi.org/10.1007/11744023_32
3. Bookstein, A., Klein, S.T.: Compression of correlated bit-vectors. Inf. Syst. **16**(4), 387–400 (1991)
4. Burrows, M. and Wheeler, D.J.: A block sorting lossless data compression algorithm. Technical report, Digital Equipment Corporation, SRC-RR-124:1–18 (1994)
5. Calonder, M., Lepetit, V., Strecha, C., Fua, P.: BRIEF: binary robust independent elementary features. In: Daniilidis, K., Maragos, P., Paragios, N. (eds.) ECCV 2010. LNCS, vol. 6314, pp. 778–792. Springer, Heidelberg (2010). https://doi.org/10.1007/978-3-642-15561-1_56
6. Choueka, Y., Fraenkel, A.S., Klein, S.T., Segal, E.: Improved hierarchical bit-vector compression in document retrieval systems. In: SIGIR 1986, Proceedings of the 9th Annual International ACM SIGIR Conference on Research and Development in Information Retrieval, Pisa, Italy, 8–10 September 1986, pp. 88–96 (1986)
7. Claude, F., Nicholson, P.K., Seco, D.: Differentially encoded search trees. In: 2012 Data Compression Conference, pp. 357–366 (2012)
8. Fraenkel, A.S., Klein, S.T.: Novel compression of sparse bit-strings – preliminary report. In: Apostolico, A., Galil, Z. (eds.) Combinatorial Algorithms on Words. NATO ASI Series (Series F: Computer and Systems Sciences), vol. 12, pp. 169–183. Springer, Heidelberg (1985). https://doi.org/10.1007/978-3-642-82456-2_12
9. Fukunaga, K., Narendra, P.M.: A branch and bound algorithms for computing *k*-nearest neighbors. IEEE Trans. Comput. **24**(7), 750–753 (1975)
10. Gálvez-López, D., Tardós, J.D.: Bags of binary words for fast place recognition in image sequences. IEEE Trans. Robot. **28**(5), 1188–1197 (2012)

11. Klein, S.T., Shapira, D.: Compressed pattern matching in JPEG images. Int. J. Found. Comput. Sci. **17**(6), 1297–1306 (2006)
12. Klein, S.T., Shapira, D.: Compressed matching for feature vectors. Theor. Comput. Sci. **638**, 52–62 (2016)
13. Leutenegger, S., Chli, M., Siegwart, R.Y.: BRISK: binary robust invariant scalable keypoints. In: 2011 IEEE International Conference on Computer Vision (ICCV), pp. 2548–2555. IEEE (2011)
14. Lowe, D.G.: Distinctive image features from scale-invariant keypoints. Int. J. Comput. Vis. **60**(2), 91–110 (2004)
15. Muja, M., Lowe, D.G.: Fast approximate nearest neighbors with automatic algorithm configuration. In: VISAPP International Conference on Computer Vision Theory and Applications, pp. 331–340 (2009)
16. Mur-Artal, R., Montiel, J.M.M., Tardos, J.D.: ORB-SLAM: a versatile and accurate monocular slam system. IEEE Trans. Robot. **31**(5), 1147–1163 (2015)
17. Nistér, D., Stewénius, H.: Scalable recognition with a vocabulary tree. In: 2006 IEEE Computer Society Conference on Computer Vision and Pattern Recognition, CVPR 2006, New York, NY, USA, 17–22 June 2006, pp. 2161–2168 (2006)
18. Rokach, L., Maimon, O.: Clustering methods. In: Maimon, O., Rokach, L. (eds.) Data Mining and Knowledge Discovery Handbook, pp. 321–352. Springer, Heidelberg (2005). https://doi.org/10.1007/0-387-25465-X_15
19. Rublee, E., Rabaud, V., Konolige, K., Bradski, G.: ORB: an efficient alternative to SIFT or SURF. In: IEEE International Conference on Computer Vision, ICCV 2011, Barcelona, Spain, 6–13 November 2011, pp. 2564–2571 (2011)
20. Trzcinski, T., Lepetit, V., Fua, P.: Thick boundaries in binary space and their influence on nearest-neighbor search. Pattern Recogn. Lett. **33**(16), 2173–2180 (2012)

D-MASC: A Novel Search Strategy for Detecting Regions of Interest in Linear Parameter Space

Daniyal Kazempour$^{(\boxtimes)}$, Kevin Bein, Peer Kröger, and Thomas Seidl

Ludwig-Maximilians-Universität München, Munich, Germany
{kazempour,kroeger,seidl}@dbs.ifi.lmu.de

Abstract. The parameter space transform has been utilized over decades in context of edge detection in the computer vision domain. However the usage of the parameter space transform in context of clustering is a more recent application with the purpose of detecting (hyper)linear correlated clusters. The runtime for detecting edges or hyperlinear correlations can be very high. The contribution of our work is to provide a novel search strategy in order to accelerate the detection of regions of interest in parameter space serving as a foundation for faster detection of edges and linear correlated clusters.

Keywords: Parameter space · Clustering · Hough transform

1 Introduction

When the parameter space transform (also known as Hough transform) was first introduced by Hough as a patent in 1962 [1] he initially designed the method with the detection of edges on electron micrograph images in mind. In the beginning of the 21st century the parameter space transform came to its application in context of high-dimensional correlation clustering with the purpose to detect (hyper)linear correlated clusters in data. This can be seen today in the CASH algorithm [5]. Although the parameter transform is a well established method in computer vision it is quite new in the field of data mining. Especially in this domain the currently available method CASH [5] reveals high runtime in certain scenarios. Further the scan of the parameter space for regions of interest (ROI) as described in [5] can in worst case deteriorate to exponential runtime. Although this is only for the worst case (we shall elaborate on the circumstances under which this case is given), it stimulates the question if there are other scan strategies to achieve faster detection of ROIs. Therefore we propose **D**ataspace **M**ean **S**hift **A**pplied **S**can for linear **C**orrelations (D-MASC). It is a heuristic approach for detecting ROIs which are further refined. The remainder of this work is structured as follows: First we sketch a brief history of the progresses made in parameter transform based methods. Then we introduce a novel scan strategy for detecting ROIs in the parameter space. We conduct experiments in which we

© Springer Nature Switzerland AG 2018
S. Marchand-Maillet et al. (Eds.): SISAP 2018, LNCS 11223, pp. 163–176, 2018.
https://doi.org/10.1007/978-3-030-02224-2_13

compare the original search strategy as used in CASH [5] to our heuristic with focus on runtime under different noise levels and numbers of clusters. Based on the experimental results we elaborate on the strengths and weaknesses of our method. We conclude by highlighting the scenarios under which our heuristic can be best utilized and provide potential targets for future work.

2   Related Work

In order to provide a solid foundation we first give a brief introduction to parameter space transform. Given a data point $p_i \in \mathbb{R}^2$. Further we are given a linear function represented for example as $b = y - mx$ where m is the slope and b the intercept of a line. While our two dimensional data space is spanned by the axis x and y, our parameter space $\mathbb{P}^2$ is spanned by the parameters of our linear function, namely m and b. A data point p_i is transformed into a data point function $b = y_{p_i} - mx_{p_i}$, representing all lines going through p_i at all possible slopes and intercepts in data space. A point $\phi_j = (m_j, b_j)$ in parameter space represents a line in data space. This duality can be seen in Fig. 1. If we have now several data points, for example p_1, p_2, p_3, we would transform each of the data points into parameter space. If we find now an intersection of the data point functions $f_{p_1}, f_{p_2}, f_{p_3}$ in parameter space at a particular point (m_s, b_s), it means that there exists a linear function in data space $y = m_s x + b_s$ on which all three points are located and therefore a linear correlation between all three data points exists as it can be seen in Fig. 2. Considering the fact the the points are not perfectly located on a line but located with an amount of jitter around the line, we are not looking for a specific point but for a region in parameter space, which we name with the term region of interest (ROI).

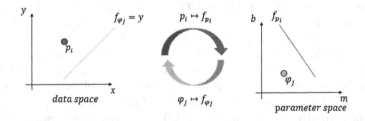

Fig. 1. Duality of data space and parameter space.

For the edge detection on electron micrographs Hough used the slope-intercept representation of lines as described before. This representation comes with several drawbacks such as unboundedness of the parameter space and the difficulties to deal with vertical lines. In order to anticipate the afore mentioned issues Duda and Hart presented in [2] a polar normal-form representation of the parameter space where a data point function is represented by

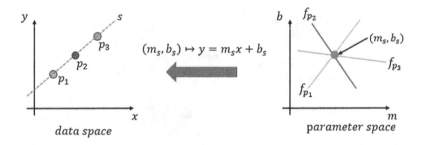

Fig. 2. An intersecting point in parameter space corresponds to a linear correlation between the involved data points in data space.

$\delta = x \cdot cos(\alpha) + y \cdot sin(\alpha)$. Here δ represents the distance from the origin perpendicular to the line and α describes the angle between an axis in parameter space and the perpendicular. Ballard further advanced the concept of parameter transform in [3] to detect arbitrary shapes. Until then the parameter space transform has been used only in the scope of edge detection on images. The very first application of the parameter space transform in the domain of data mining was conducted by Achtert et al. in [5]. Here the authors introduced several novel aspects which emerged from the necessity to deal with high-dimensional data. The purpose of using parameter space transform in their work, is to detect arbitrary oriented hyper-linear correlated clusters.

3   Search Strategies

We now proceed to give a brief overview on the first methods for detecting ROIs in parameter space. Then we describe the original strategy for scanning the parameter space for ROIs as described in [5], and continue to elaborate on our developed heuristic.

3.1   The History of Parameter Space Search Strategies

The original parameter space algorithms in pattern recognition on image data had a voting system based on the idea that the parameter space is segmented into equi-sized squares, just like applying a static grid on the parameter space. These squares are named accumulators and the accumulators with the highest counts (referred to as votes) of functions intersecting are potential ROIs as seen in Fig. 3. This approach however comes with a major drawback: If the original grid consists of $n \times n$ accumulators, with every detected high-voted accumulator, another $n \times n$ grid is created. Given the case that many accumulators have a high vote, as a consequence the number of $n \times n$ accumulators increases which have to be checked for their respective vote count. Considering high-dimensional data sets, the number of hyper-accumulators grows exponentially by dimension d.

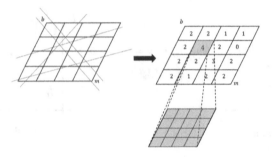

Fig. 3. Accumulator based approach with voting.

The authors in [4] proposed a hybrid approach in which they use a pre-defined grid of a coarse resolution. If any of the coarse grid accumulators are sufficiently voted for, the affected accumulator is split among a pre-defined order among the axis of the parameter space into half in a kd-tree like fashion as seen in Fig. 4, where those halves which have sufficient support but are not the highest so far, are put into a queue.

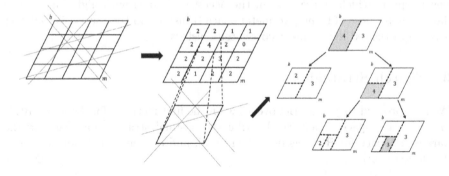

Fig. 4. Hybrid accumulator and successive splits based approach.

The concept of successively splitting the parameter space among a pre-defined axis-order has been utilized in [5] in context of detecting arbitrary oriented hyperliniear correlated clusters. However, this method can still become significantly slow. Especially in cases where the data set contains linear correlations with a high amount of jitter or a high amount of noise it can lead to many small candidate accumulator cells as it can be seen in an example in Fig. 5.

We use the polar representation in our work where each data point is represented as a sinusoidal curve in parameter space. This representations has some advantages over the slope-intercept representation, such as its parameter space is bounded where the slope-intercept representation is unbound.

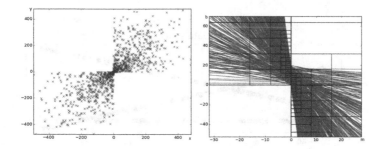

Fig. 5. Synthetic data sets. left: scattered line, right: parameter space with a large number of candidate cells

3.2   Data Space Prediction

Our concept in D-MASC relies on the core idea of sampling the data points from data space. The resulting data set will be smaller in size, leading to a smaller number of functions in parameter space. To achieve this the Mean Shift [6] algorithm is used, where each data point roams to its mode with respect to its neighboring data points within a specific window, named in literature as bandwidth. Mean Shift is applied here on the data space instead of the parameter space. The steps of D-MASC are described below and can be seen in Fig. 6.

1. Given a data set DB. Apply the Mean Shift algorithm on the original set.

   $\Gamma = \mathtt{mean\_shift}(\mathrm{DB})$

2. Convert the obtained modes from Mean Shift to parameter space functions.

   $\mathcal{P} = \mathrm{map}(\Gamma, \mathtt{toParamSpace})$

3. Sample the parameter functions along their paths and fix a search area around each sample in such a way that no gaps emerge. This concept is like a rasterization of the mode functions.
4. Count the number of intersections in each search area and determine whether it satisfies the minimum point condition. If it does, a cluster is found.

   (1) It might be beneficial to select a sample of the set to further decrease the computation time. In order retain the correctness and precision of the prediction, the set must be sufficiently big. Sampling small sets will result in a significant information loss and the calculated modes may be off. In its intended version, Mean Shift is targeted to find "blob" like cluster structures (using a Flat or Gaussian Kernel). To a certain extend, this also applies to linear clusters. Usually, a line in data space contains segments that are more densely aligned than its surrounding noise or other parts of that line. These segments are less linearly correlated than the cluster itself and thus Mean Shift will tend to converge against them.

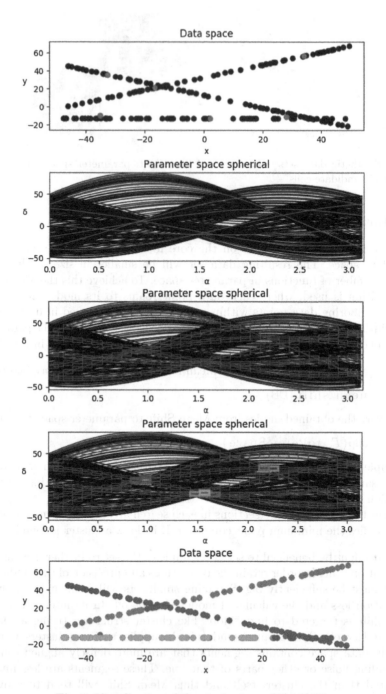

Fig. 6. Visualization of the single steps of D-MASC

(2) The modes must then be parameterized, which creates a mode function for each mode in parameter space. It is important to notice that this mode function is defined in the whole interval $[0, \pi]$. Since the mode should be very close to a linear cluster, its mode function will be intersecting the clusters' data point functions at some point. To find this point, the whole functions must be scanned.

(3) Projecting the modes to the mode functions and determining the search areas is a challenging task. Seeing that a cluster can be located on any point on the mode function, the search areas around the functions shall not have gaps between them. There are two variables that need to be picked: The number of sampling points `sample_count` and the height h of the search areas. The number of sampling points affects the performance and the result. Choosing too few will return inaccurate clusters. On the contrary, choosing to many will drastically increase the computation time. To be consistent with and comparable to the original strategy, the `sample_count` is calculated depending on the `max_level`:

$$\texttt{sample\_count} = \pi \cdot (0.5^{\beta}) \text{ with } \beta = \left\lceil \frac{\texttt{max\_level}}{2} \right\rceil$$

The width w is directly linked to the number of sampling points `sample_count`. The goal is to have no overlapping areas and no (horizontal) gaps between the areas. w can be calculated as follows:

$$w = \frac{1}{\texttt{sample\_count}}$$

The height h can also be prone to errors. If the height is too small, the data point function might not be sufficiently covered or leave (vertical) gaps. This is precisely what happens when h is calculated based on the `max_level`, analogously to above as can be seen in Fig. 7.

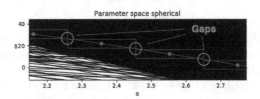

Fig. 7. Vertical gaps between search areas when the height is too small.

Instead, testings suggested the following series s to be adequate height factors (other series may also be possible):

$$h = (\delta_{\max} - \delta_{\min}) \cdot s_{\texttt{max\_level}} \text{ with } s_i = 1, \frac{3}{4}, \frac{2}{4}, \frac{3}{8}, \frac{2}{8}, \frac{3}{16}, \frac{2}{16}, \frac{3}{32}, \frac{2}{32}, \frac{3}{64}, \cdots$$

The search areas Λ_i can now be defined based on the previous definitions and calculated modes (cx_i, cy_i) (see Fig. 8):

$$\Lambda_i = \{((cx_i - \frac{w}{2}, \ cy_i - \frac{h}{2}), \ w, \ h)$$

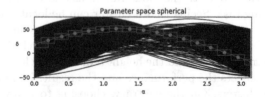

Fig. 8. Search areas Λ_i around the samples of a parameterized cluster found with Mean Shift.

This technique eliminates the need to look at sparse locations (usually the upper and lower area of the whole space and "holes" between paths). This strategy is not suited for finding clusters with a high resolution like the original strategy is capable of, with a high `max_level`. We shall see in the following section that choosing a large number of intersections s will have a large impact on the runtime of D-MASC which also will be further confirmed by the tests conducted in the experiment section.

3.3 Complexity

In this section we elaborate on the runtime complexity of D-MASC and oppose it to the runtime complexity of the original strategy as proposed in CASH. D-MASC consists of two steps with their respective complexities:

- Computing modes in data space using Mean Shift: The runtime complexity of Mean Shift is $\mathcal{O}(Tn \log n)$ where T is the number of iterations and n denotes the number of data points. The number of iterations can be influenced through the chosen bandwidth hyperparameter of Mean Shift.
- Voting: For all the bounding boxes of all mode functions it is counted how many mode functions intersect a particular bounding box. The number of checks per box is the number of mode functions m leading to a runtime complexity of $\mathcal{O}(\frac{len(bounds_d)}{w} mm)$, where the width w is defined as $w = \frac{1}{\frac{\pi}{2}\frac{s}{2}}$. Here the number of intersections s is included, where a large value has a high resolution of the corresponding boxes and a small value a low resolution.

All the above mentioned runtime complexities result in the overall runtime complexity of $\mathcal{O}(Tn \log n + \frac{len(bounds_d)}{w} m^2)$. The runtime complexity of CASH can deteriorate to $\mathcal{O}(s \cdot c \cdot n \cdot 2^d)$ where c is the number of resulting clusters. This worst case emerges especially if the data set has a high level of noise, as we shall see in the experiments or the clusters have a high amount of jitter. The runtime complexity of D-MASC with regards to higher dimensional data is: $\mathcal{O}(Tn \log n + (\frac{len(bounds_d)}{w})^d m^2)$.

In order to get an intuition under which circumstances D-MASC excels we have conducted the following computation to better comprehend the behavior of CASH and D-MASC. For ensuring comparability we set in both scenarios the dimensionality $d = 8$ and the number of intersections $s = 10$ equal for both methods. Further we used data set sizes of $n = 10^{15}, 10^{14}$ and $n = 10^{13}$ for both Algorithms. As it can be seen in Fig. 9, with increasing number of clusters found the runtime of cash grows up linearly to 10^{19} computation steps, while D-MASC achieves on the same amount of data points, by a factor of 10 a number of computation steps of 10^{18}. The number of modes have been set to $m = 1, 8, 16$ with increasing number of data points and the number of iterations was kept constant at $T = 10$. We shall see in more detail under which circumstances D-MASC outperforms the original CASH strategy in 2D settings.

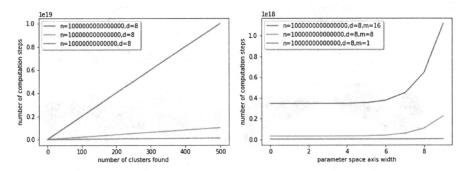

Fig. 9. Left: number of computation steps with increasing number of clusters found in CASH; Right: number of computation steps with increasing parameter space axis width. All experiments conducted at the same three different data set sizes n and with the dimensionality $d = 8$.

4   Experiments

Having elaborated on the scan strategy for ROIs in parameter space, we now proceed to evaluate the performance of D-MASC. For this we first describe briefly the setup, and present the used data sets. We then compare our method against the original scan strategy as used in CASH regarding different noise levels and sizes of data as well as regarding different hyperparameter settings. We focus in this paper on two-dimensional data sets as our primary goal is to check the performance of D-MASC regarding different noise levels and sizes of data. Further we relinquish the usage of real world data, since the artificial data sets reflect larger volumes as well as a broad range of noise levels. The tests were

run on a system with an Intel(R) Core(TM) i7-3740QM @ 2.7GHz, 4 Cores and
16.0 GB of physical memory whereas each test run received one single thread. No
parallelism was applied to the algorithms in order to retain comparable results.

4.1   Data Sets

In the following experiments a couple of data sets containing linear clusters are
used. Two basic artificial data sets are generated with 1 000, 10 000 and 100 000
points each. The noise levels range from 0% to 90%. To provide meaningful
correlations for the algorithms to find, the first set contains one linear cluster,
which goes from middle left to top right in the upper half. The second set contains
three linear clusters with one being parallel to the x-axis, one going from top left
to bottom right and one going from bottom left to top right. Noise is randomly
distributed around each of the clusters. Figure 10 shows a preview of the data
sets.

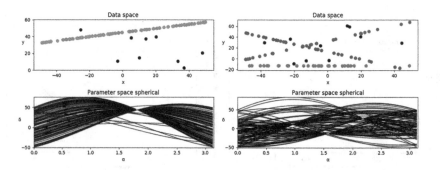

Fig. 10. Preview of the two basic test data sets. The plot on the left side contains 1
cluster (orange) and the plot on the right contains 3 clusters (green, red, blue). Noise
(black, here 10% only) is distributed around them. (Color figure online)

These sets were generated before any tests were conducted. The `min_count`
differed between the sets and depended on the number of clusters and the amount
of noise. Table 1 shows the `min_count` that was used for each data set. The values depend on the maximum number of points per cluster. The general intuition
of `min_count` is that if the parameter is chosen with a high value, the resulting clusters may be in numbers few, but each cluster contains a high (at least
`min_count`) number of data points, thus giving a higher confidence of being a
linear correlated cluster.

4.2   Data Space Prediction

As an initial idea we have performed a sampling of the data point functions in
parameter space and have also applied Mean Shift in parameter space. However the D-MASC approach which *first* applies Mean Shift in *data space* always

Table 1. Values of min_count for each data set.

	1 linear			3 linear		
	1 000	10 000	100 000	1 00	10 000	100 000
0%	1 000	10 000	100 000	330	3 300	33 000
10%	900	9 000	90 000	300	3 000	30 000
50%	500	5 000	50 000	160	1 600	16 000
75%	250	2 500	25 000	83	830	8 300
90%	100	1 000	10 000	33	330	3 300

returned more precise cluster results than the approach of applying Mean Shift on sampled data point functions in parameter space. D-MASC even works on data sets with mixed cluster structures (perfectly linear and more scattered lines which can as well be regarded as blobs) where the parameter space approach failed as it can be seen in Fig. 11.

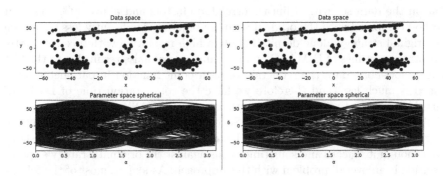

Fig. 11. Comparison between detecting regions of interest by applying Mean Shift on sampled data point functions in parameter space (left side) and detecting ROIs by using D-MASC(right side) with set mixed_1000_50. This data set contains 3 blob like structures in the bottom half and a line in the top half.

Under certain conditions, D-MASC has been observed to have a significantly better performance than the original strategy as used in CASH. If it is possible for the prediction stage to outperform the original algorithm up to a certain level while maintaining a relatively high precision at the same time, D-MASC will be faster. By afterwards refining the predicted areas using the original strategy there will be no additional loss in performance.

The two key parameters are the max_level (also known as s) and the bandwidth for the Mean Shift algorithm. We have empirically determined that on our used data sets a bandwidth value of around 0.2 gives the best trade-off between precision and runtime performance for the test data sets. Selecting the max_level and thus the size of the search areas around the mode functions is critical regarding the resolution of the results. To briefly recap D-MASC consist of three tasks: Applying Mean Shift in data space to obtain modes, fixing search areas around the resulting mode functions in parameter space and calculating the number of intersections in the search areas. As the graphs in Fig. 12 show, our D-MASC strategy works especially well with noisy data sets and sets with multiple clusters. Still, the original strategy will be superior in sets where close to no noise is present.

By analyzing Fig. 12, it can be deduced that a speedup is heavily correlated with the used data set. Running the strategy on sets with high noise levels above 50% already achieves improvements with a `max_level` starting as early as 4 or 5. Under such circumstances, D-MASC performs with lower runtimes compared to the original strategy. The problem with low maximum levels is the precision. Although the speedup is increasing, the precision starts declining radically. At `max_level` ≥ 9 the strategy already fails in our experiments to identify the clusters correctly. If a higher resolution is desired, the original strategy has to be applied from level 9 onward to yield the expected results. This hybrid approach of using D-MASC first and on the resulting candidates to apply the original CASH strategy proved to a good solution. For example in Table 2 it can be seen that on the data set with 3 linear correlated clusters and above 50% noise our method combined with the original strategy requires significantly less time then the original strategy alone despite having a high resolution with `max_level=16`. The larger runtimes of D-MASC on lower amounts of noise result due to the fact that on low-noise data the reduction of data point function in parameter space gained is much less, and therefore we have the computation time of D-MASC plus the additional refinement computation time of CASH.

Hence, an upper bound for D-MASC on the test data sets is `max_level` $= 8$. Everything below will miss obvious clusters as the previous experiments show. Even though it reaches and sometimes wins against the original strategy's speed, precision is an overall problem with this approach. As long as most of the following aspects apply to the data set, it will perform better regarding the runtime aspect: The data set has a high noise or a high jitter level, contains a large number of clusters and the desired maximum level and thus the resolution is low.

Table 2. Left: test results for the original strategy, right: test results for D-MASC combined with using the orignal strategy for refinement. `max_level` $= 16$.

	3 Linear orig.			3 Linear D-MASC		
	1 000	10 000	100 000	100	10 000	100 000
0%	1.27 s	12.83 s	128.54 s	3.36 s	32.67 s	326.82 s
10%	1.32 s	13.62 s	136.28 s	3.29 s	31.94 s	334.88 s
50%	2.02 s	20.57 s	232.74 s	2.47 s	26.17 s	181.13 s
90%	8.79 s	90.17 s	1065.26 s	5.16 s	35.56 s	564.35 s

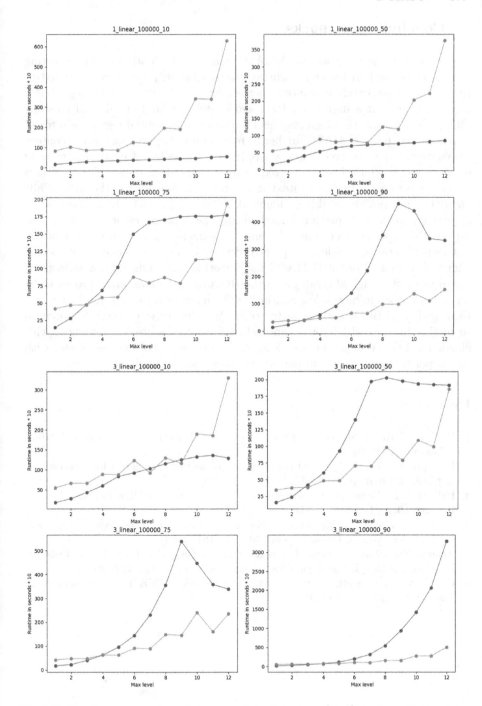

Fig. 12. Runtime comparison between original strategy (blue) and predicting data space (orange) using data sets 1_linear_100000_* and 3_linear_100000_*. (Color figure online)

5   Concluding Remarks

In our work we have presented D-MASC, a novel search strategy for detecting regions of interest in linear parameter space. This strategy is very limited to specific cases yet excels in runtime performance if the data set is large, has a high noise level or a high jitter level and contains a large number of clusters. Further it is a heuristic which may lead to the loss of potential regions of interest in parameter space. D-MASC further introduces novel approaches in the context of parameter space search like e.g. applying a mode based clustering method on data space which leads to a reduction of candidate data point functions in parameter space. In this context it is vital to highlight that the Mean Shift algorithm may not be suitable for higher dimensional data sets. Here more robust approaches such as Laplacian k-medoid may be a better choice. Beyond that aspect, D-MASC does not split the entire parameter space into smaller boxes, but rather rasterizes single data point functions into smaller boxes with specific width. A combination of D-MASC for detecting the ROIs with a refinement phase using the original strategy on the ROIs in parameter space proves to be a good approach to have a low runtime while maintaining a high resolution and thus quality of the resulting cluster cells. As a future work there is potential to further improve the hybrid approach, eventually by also combining it with classic fixed-grid voting schemes. In general we hope to motivate and foster with this paper further research in parameter space scan strategies.

References

1. Paul, V.C.: Hough: method and means for recognizing complex patterns. US Patent US3069654A (1960)
2. Duda, R.O., Hart, P.E.: Use of the Hough transformation to detect lines and curves in pictures. Commun. ACM **15**(1), 11–15 (1972)
3. Ballard, D.H.: Generalizing the Hough transform to detect arbitrary shapes. Pattern Recogn. **13**(2), 111–122 (1981)
4. Li, H., Lavin, M.A., Le Master, R.J.: Fast Hough transform: a hierarchical approach. Comput. Vis. Graph. Image Process. **36**, 139–161 (1986)
5. Achtert, E., Böhm, C., David, J., Kröger, P., Zimek, A.: Global correlation clustering based on the Hough transform. Stat. Anal. Data Min. **1**, 111–127 (2008)
6. Cheng, Y.: Mean shift, mode seeking, and clustering. IEEE Trans. Pattern Anal. Mach. Intell. **17**, 790–799 (1995)

On the Correlation Between Local Intrinsic Dimensionality and Outlierness

Michael E. Houle[1], Erich Schubert[2(✉)] [iD], and Arthur Zimek[3] [iD]

[1] National Institute of Informatics,
2-1-2 Hitotsubashi, Chiyoda-ku, Tokyo 101-8430, Japan
meh@nii.ac.jp
[2] Heidelberg University, Im Neuenheimer Feld 205, 69120 Heidelberg, Germany
schubert@infomatik.uni-heidelberg.de
[3] University of Southern Denmark, Campusvej 55, 5230 Odense M, Denmark
zimek@imada.sdu.dk

Abstract. Data mining methods for outlier detection are usually based on non-parametric density estimates in various variations. Here we argue for the use of local intrinsic dimensionality as a measure of outlierness and demonstrate empirically that it is a meaningful alternative and complement to classic methods.

Keywords: Outlier detection · Intrinsic dimensionality · Comparison

1 Introduction

The classic statistics literature on outlier detection provides some definitions of outliers as follows:

> "The intuitive definition of an outlier would be 'an observation which deviates so much from other observations as to arouse suspicions that it was generated by a different mechanism'." – Hawkins [20]

> "An outlying observation, or 'outlier,' is one that appears to deviate markedly from other members of the sample in which it occurs." – Grubbs [17]

> "An observation (or subset of observations) which appears to be inconsistent with the remainder of that set of data." – Barnett and Lewis [5]

These classic albeit rather informal definitions of outlierness based on vague descriptions like "arouse suspicions", "appear to deviate", or "appear to be inconsistent" highlight that outlierness is a rather inexact property. Consequently, there are many different techniques to identify outliers. However, most of them are based on some notion of likelihood, typically based on the probability density distribution that is estimated locally from the data sample at

© Springer Nature Switzerland AG 2018
S. Marchand-Maillet et al. (Eds.): SISAP 2018, LNCS 11223, pp. 177–191, 2018.
https://doi.org/10.1007/978-3-030-02224-2_14

hand. Potential outliers are then those objects located in less dense areas of some dataset, be it on an absolute (global) or a relative (local) scale.

The local intrinsic dimensionality (LID) can be seen as a different approach to capture the "inconsistency" of outliers, not being based on the density distribution in the dataset but rather on the distribution of distances. Although both views are related, LID is a genuinely different technique to capture outlierness. In this paper, we argue theoretically and practically for the usefulness of local intrinsic dimensionality as a measure for outlierness.

In the remainder, we survey related work regarding outlier detection, (local) intrinsic dimensionality, and the estimation of local intrinsic dimensionality (Sect. 2), we present some arguments for the relation between local intrinsic dimensionality and outlierness (Sect. 3), and we present an empirical evaluation of the suitability of local intrinsic dimensionality as a measure of outlierness (Sect. 4). We conclude with perspectives for future research (Sect. 5).

2   Related Work

2.1   Outlier Detection

Many outlier detection methods are available in the literature. They differ in the way they model and identify outliers and, thus, in the assumptions they, implicitly or explicitly, rely on.

Statistical methods for outlier detection (also: outlier identification or rejection) are typically based on assumptions on the nature of the distributions of objects. The classical textbook of Barnett and Lewis [5] discusses numerous tests for different distributions. These tests are optimized for each distribution, depending on the specific parameters of the corresponding distribution, the number of expected outliers, and the location where to expect an outlier. Various statistical techniques have been discussed by Rousseeuw and Hubert [52].

The first database-oriented approach by Knorr and Ng [37] was inspired by a statistical model but approaches the problem as non-parametric learning, i.e., not assuming specific data distributions. Their method performs local density estimates, counting the number of objects within an ε-range around each point. Points with too small neighborhoods (as given by the ε threshold) are qualified as outliers (i.e., the local density is too low). Instead of ε ranges, the k-nearest neighbor outlier model [50] uses the distances to the kth nearest neighbor (kNN) of each object to rank the objects. As a variant, the sum of distances to all points within the set of k nearest neighbors (called the "weight") has been used as an outlier degree [4].

The so-called "local" approaches consider ratios between the local density around an object and the local density around its neighboring objects, starting with the seminal LOF [7] algorithm, based on the idea that "normal" density may be different in different regions of the dataset. Many variants adapted the original LOF idea in different aspects [57].

A common source of variation is the definition of neighborhoods using, e.g., approximations [13,48,58,61], 'connectivity' [60], clusters [9], or reverse neighborhoods [19,34,49]. LOCI [47] uses ε-neighborhoods rather than kNN and integrates over varying ε values.

Other methods vary the way local estimates are put in relation to each other (model) or the way, the density is actually estimated. LDOF [62] considers the average pairwise distances within the set of k nearest neighbors. LoOP [38] puts the focus on a probabilistic interpretation of outlier scores. LDF [42] and KDEOS [56] redefine the density estimation as a kernel density estimator. Angle-based outlier detection (ABOD) [40] assesses the variance of pairwise angles between all triples of points.

Many of these methods have been evaluated by Campos et al. [10], along with a study of 23 base datasets in various variants, resulting in more than 1000 different datasets. We will compare local intrinsic dimensionality as an outlier estimator against these benchmark results that are available online.[1]

2.2   Local Intrinsic Dimensionality

Over the past decades, many characterizations of intrinsic dimensionality have been proposed. The earliest theoretical measures of ID, such as the classical Hausdorff dimension, associate a non-negative real number to sets in terms of their covering or packing properties with respect to an underlying metric space (for a general reference, see [14]). These theoretical measures have served as the foundation of practical methods for finite data samples that estimate ID in various ways: from the space-filling capacity or self-similarity properties of the data [8,16,18], as the basis dimension of the tangent space of a data manifold from local samples (see for example [53]); as a byproduct when determining lower-dimensional projective spaces or approximating surfaces (such as with principal component analysis [35]); or in parametric modeling and estimation of distribution [41,43].

Classical expansion models of dimensionality, such as the expansion dimension, quantify the ID in the vicinity of a point of interest in the data domain, by assessing the rate of growth in the number of data objects encountered as the distance from the point increases. Expansion models take their motivation from the observation that in many metric spaces (including Euclidean space), the volume of an m-dimensional ball grows proportionally to r^m as its radius is scaled by a factor of r. From this rate of volume growth with distance, the dimension m can be deduced as:

$$\frac{V_2}{V_1} = \left(\frac{r_2}{r_1}\right)^m \Rightarrow m = \frac{\ln(V_2/V_1)}{\ln(r_2/r_1)}. \tag{1}$$

This expression of dimension in terms of the radius and volume of a neighborhood ball is exploited by classical expansion models, by treating the number of data points captured by a ball as a proxy for its volume [26,36,46,53].

---

[1] http://www.dbs.ifi.lmu.de/research/outlier-evaluation.

More recently, a distributional form of intrinsic dimensional modeling was proposed, the Local Intrinsic Dimensionality (LID) [23, 24] in which the volume of a ball of radius r is taken to be the probability measure associated with its interior, denoted by $F(r)$. The function F can be regarded as the cumulative distribution function (cdf) of an underlying distribution of distances. Adapting Eq. 1, intrinsic dimensionality can then be modeled as a function of the distance r, by letting the radii of the two balls be $r_1 = r$ and $r_2 = (1 + \epsilon)r$, and letting $\epsilon \to 0$. The following definition (as stated in [24]) generalizes this notion even further, to any real-valued function that is non-zero in the vicinity of $r \neq 0$.

Definition 1 ([24]). *Let F be a real-valued function that is non-zero over some open interval containing $r \in \mathbb{R}$, $r \neq 0$. The* intrinsic dimensionality *of F at r is defined as follows whenever the limit exists:*

$$\text{IntrDim}_F(r) \triangleq \lim_{\epsilon \to 0} \frac{\ln\left(F((1+\epsilon)r)/F(r)\right)}{\ln\left((1+\epsilon)r/r\right)} = \lim_{\epsilon \to 0} \frac{\ln\left(F((1+\epsilon)r)/F(r)\right)}{\ln(1+\epsilon)}.$$

Under the same assumptions on F, where $F(r)$ is associated with the probability of a random distance variable being less than r, the definition of LID can be shown to be equivalent to a notion of indiscriminability of the distance measure. More precisely, if F is discriminative at a given distance r, then expanding the distance by some small factor should incur a small increase in probability measure as a proportion of the value of $F(r)$. Conversely, if F is indiscriminative at distance r, then the proportional increase in probability measure would be large. Accordingly, the indiscriminability of the distance variable is defined as the limit of the ratio of two quantities: the proportional rate of increase of probability measure, and the proportional rate of increase in distance.

Definition 2 ([24]). *Let F be a real-valued function that is non-zero over some open interval containing $r \in \mathbb{R}$, $r \neq 0$. The* indiscriminability *of F at r is defined as follows whenever the limit exists:*

$$\text{InDiscr}_F(r) \triangleq \lim_{\epsilon \to 0} \left[\frac{F((1+\epsilon)r) - F(r)}{F(r)} \middle/ \frac{(1+\epsilon)r - r}{r} \right] = \lim_{\epsilon \to 0} \frac{F((1+\epsilon)r) - F(r)}{\epsilon \cdot F(r)}.$$

When F satisfies certain smoothness conditions in the vicinity of r, its intrinsic dimensionality and indiscriminability have been shown to be identical:

Theorem 1 ([24]). *Let F be a real-valued function that is non-zero over some open interval containing $r \in \mathbb{R}$, $r \neq 0$. If F is continuously differentiable at r, then*

$$\text{ID}_F(r) \triangleq \frac{r \cdot F'(r)}{F(r)} = \text{IntrDim}_F(r) = \text{InDiscr}_F(r).$$

As a way of characterizing the local intrinsic dimensionality in the vicinity of $\mathbf{x}$ in a way that does not depend on distance, we are interested in the limit of $\text{ID}_F(r)$ as r tends to 0. For convenience, for non-zero distances r we refer to $\text{ID}_F(r)$ as the *indiscriminability of F at distance r*, and to $\text{ID}_F^* \triangleq \lim_{r \to 0} \text{ID}_F(r)$ as the *local intrinsic dimension of F*.

Let $\mathbf{x}$ be any reference location within a data domain $\mathcal{D}$ equipped with a distance measure d. To any point $\mathbf{y} \in \mathcal{D}$ we can associate the distance $r = d(\mathbf{x}, \mathbf{y})$; in this way, any global data distribution over $\mathcal{D}$ induces a local distance distribution with respect to $\mathbf{x}$. Given a sample of n points drawn from the global distribution, the cdf of the local distance distribution can be represented as a function F as described above. Given a neighborhood radius r, the expected number of points from a sample of size n that lie within the neighborhood is simply $n \cdot F(r)$. Conversely, the k-nearest neighbor distance within a sample of n points is an estimate of the distance value r for which $F(r) = k/n$. If F is smooth, k is fixed, and n is allowed to tend to infinity, the indiscriminability of F at the k-nearest neighbor distance tends to the local intrinsic dimension. The local intrinsic dimension can thus serve to characterize the degree of difficulty in performing similarity-based operations within query neighborhoods using the underlying distance measure, asymptotically as the sample size (that is, the dataset size) scales to infinity.

In the ideal case where the data in the vicinity of $\mathbf{x}$ is distributed uniformly within a submanifold in $\mathcal{D}$, ID_F^* would equal the dimension of the submanifold; however, in general these distributions are not ideal, the manifold model of data does not perfectly apply, and ID_F^* is not necessarily an integer. Nevertheless, the local intrinsic dimensionality would give a rough indication of the dimension of the submanifold containing $\mathbf{x}$ that would best fit the data distribution in the vicinity of $\mathbf{x}$. We refer readers to [24,25] for details concerning the LID model.

2.3   Estimators of Local Intrinsic Dimensionality

A strong connection has been shown between local intrinsic dimensionality and the scale parameter of the generalized Pareto distribution [2], and to the index of regularly varying functions in the Karamata representation from extreme value theory [24]. In [2], an extreme-value-theoretic framework was used to derive several estimators of ID_F^*, using the techniques of maximum likelihood estimation (MLE), the method of moments, and probability weighted moments. The MLE estimator derived for LID was shown to coincide with the classical Hill estimator developed within the extreme value theory research community [22]:

$$\widehat{\text{ID}}_F^* = -\left(\frac{1}{k} \sum_{i=1}^{k} \log \frac{r_i}{r_k}\right)^{-1}, \tag{2}$$

where $(r_i \mid 1 \leq i \leq k)$ is a sample of k distance measurements associated with a k-nearest neighbor set, taken in non-decreasing order. In addition, a family of estimators was also derived using the theory of regularly varying functions, three special cases of which coincided with two discrete expansion-based models of intrinsic dimensionality—the Generalized Expansion Dimension (GED) [26], and a special case of the MiND family of estimators [53]—as well as a third that tended to the MLE/Hill estimator as the sample size k tends to infinity.

In [2], an experimental study was conducted on the aforementioned estimators of LID, as well as several other measures of intrinsic dimensionality, both local and global: principal component analysis (PCA) [35], the correlation dimension [8,21,59], methods based on nearest-neighbor graphs [12], and MiND [53]. The experiments generally found that the local estimators of intrinsic dimensionality were better at identifying the intrinsic dimensionality of artificial datasets generated on non-linear (curved) manifolds, as compared to global methods. Amongst the local estimators tested, the performance of the various estimators was comparable, with the Hill/MLE estimator achieving a slightly better trade-off between convergence and bias. For this reason, in all our experimentation we chose to implement the estimation of LID using an aggregated variant of the Hill estimator proposed for smaller neighborhood sizes, due to Huisman et al. [33].

The expansion models of intrinsic dimensionality were originally conceived for the analysis of the performance of indexing techniques [6,32,36]. More recently, estimators of LID (and in particular, the Hill estimator) have been shown to be practically effective in guiding the runtime decisions of applications of similarity search, machine learning, and data mining. Examples of these include: effective early termination of similarity search using runtime tests based on intrinsic dimensionality [11,27–30]; guiding the local sparsification of features concurrently with similarity graph construction [31]; the analysis of non-functional dependencies among data features [51]; a theoretical analysis of the effect of adversarial perturbation on classification [1] and a practical deep neural network detection method for adversarial examples [44]; and a deep neural network classification strategy that monitors the dimensionality of subspaces during training and adapts the loss function so as to limit overfitting [45].

Expansion models of dimensionality hold an advantage over parametric models in that they require no explicit knowledge of the underlying global data distribution. They also have the advantage of computational efficiency: as they require only an ordered list of the neighborhood distance values, no expensive vector or matrix operations are required for the computation of estimates. Although the Hill estimator typically requires on the order of 100 neighborhood samples for convergence [2], smaller samples (even on the order of 10) have nevertheless been shown to provide useful information in the context of deep neural network classification [44,45].

3   Local Intrinsic Dimensionality and Outlierness

As discussed in Sect. 2.1, outlierness is often assessed in terms of local density; in particular, the popular LOF algorithm and its variants contrast the local density in the vicinity of a test point to the density at its neighbors. In [24] it was shown that fluctuations in density can be understood in terms of variation in the LID across the data domain. In the ideal situation where the domain is a submanifold, and the data distribution is uniform, the value of ID_F^* taken at any point in the relative interior of the submanifold would be the same—the dimension of the submanifold itself. However, when the distribution is not uniform, changes in density can be revealed by changes in the local intrinsic dimension. In [24], this was investigated in terms of a second-order theory of intrinsic dimensionality, which has its analogue in a second-order theory of extreme values originally developed to account for the rate of convergence of estimators of extreme-value-theoretic parameters. Although second-order LID can theoretically reveal changes in growth rates, estimation of these parameters within the context of second-order extreme value theory does not appear to be practical: no closed-form estimators are known, and many thousands of samples are required for the convergence of the non-closed-form estimators that do exist [15].

A second argument for the relationship between density, outlierness and local intrinsic dimensionality can be made, based on the a precise theoretical characterization of distance distributions in terms of LID; although proven from first principles in [24], it is a somewhat sharper statement of the classical Karamata representation theorem from extreme value theory, within the context of smooth distance distributions. Interpreted asymptotically as the neighborhood radii tend to zero, it essentially states that the relationship between the sizes and radii of two concentric neighborhood samples is ultimately governed by the intrinsic dimensionality. Informally, the relationship can be stated as follows (for the precise statement, please see [24]):

$$\frac{F(r_2)}{F(r_1)} \approx \left(\frac{r_2}{r_1}\right)^{\mathrm{ID}_F^*}. \tag{3}$$

Clearly, knowledge of ID_F^*, either through modeling or through estimation, reveals relationships among the expected sizes and radii of neighborhoods around points of interests—the same information used in many if not most of the state-of-the-art outlier detection methods. In this paper, rather than designing specific tests for outlier detection in terms of LID, our goal is to establish that a straightforward use of LID estimates—even with the bias and variation that are associated with parameter estimation—can reveal outlierness in a manner that is competitive with the state of the art. Since the LID model makes no assumptions on the nature of the data distribution other than smoothness (which is required for generalization anyway), and since it is naturally adaptive to local variation in intrinsic dimensionality, it has the inherent potential to allow the ranking of outlier candidates even when the dimensional characteristics are highly nonuniform.

4   Experimental Evaluation

For the experimental evaluation, we compare LID as an outlier measure to the results in the benchmark collection of Campos et al. [10], that are completely available online.[2] The competing methods in the repository are classic neighborhood-based methods, namely kNN [50], kNN-weight (kNNW) [3,4], ODIN (Outlier Detection using Indegree Number) [19], LOF (Local Outlier Factor) [7], SimplifiedLOF [57], COF (Connectivity-based Outlier Factor) [60], INFLO (Influenced Outlierness) [34], LoOP (Local Outlier Probabilities) [38],

---

[2] http://www.dbs.ifi.lmu.de/research/outlier-evaluation.

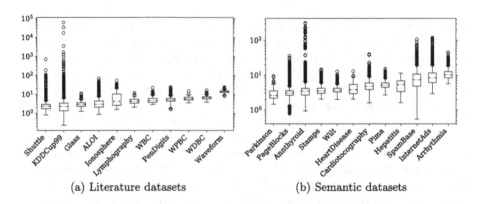

(a) Literature datasets   (b) Semantic datasets

Fig. 1. Distribution of LID estimates for different datasets.

LDOF (Local Distance-based Outlier Factor) [62], LDF (Local Density Factor) [42], KDEOS (Kernel Density Estimation Outlier Score) [56], and FastA-BOD (Fast Angle-Based Outlier Detection), a variant of angle-based outlier detection (ABOD) [40]. We additionally include ISOS [54], which is a significantly more complicated approach based on stochastic neighbor weight distribution involving intrinsic dimensionality.

We test LID, using the Hill estimator for estimating LID in the same framework ELKI [55] as the experiments in the benchmark collection [10]. We also use the same settings (ranging the neighborhood size up to 100, if the dataset is not smaller), on the same datasets, and the same performance measures.

The intrinsic dimensionality average values of all the datasets in this benchmark repository are fairly low. We plot the distribution of LID values over the different datasets (we include the normalized dataset variants without duplicates and with a range of 3–5% of outliers) in Fig. 1. While the majority of local estimates of the intrinsic dimensionality is in the order of 10, some values in each dataset are several orders of magnitude larger, hinting at outlierness.

As this benchmark repository contains more than 1000 datasets as variants of these 23 basis datasets, we resort to summary plots on the average performance for a general picture of the usefulness of local intrinsic dimensionality as a measure of outlierness. A main summary for the comparison of methods is given in Fig. 2. These plots allow to compare the average performances of the tested methods in terms of the average precision over all datasets (Fig. 2(a)), the average performance with the best parameter choice (Fig. 2(b)), and the average performance within a window of size 5 (Fig. 2(c)) and 10 (Fig. 2(d)) around the best parameter choice. For each method and setting, we report the mean and the standard error as aggregated over all datasets with various proportions of outliers (we restrict the analysis to the normalized datasets without duplicates, following Campos et al. [10]). While LID is on average worse than most of the classic methods (we would not expect a straightforward application of LID as

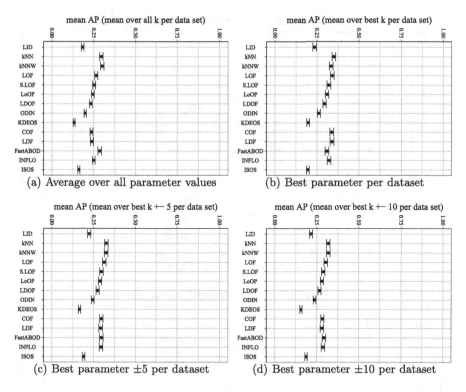

Fig. 2. Performance of methods averaged over all datasets (normalized, without duplicates).

outlier measure to beat established methods), it is still in the main field and better than some of the established methods such as KDEOS and ISOS.

While these plots show the average performance over all datasets of a certain characteristic, it should be noted that of course different methods can excel on different datasets. Also the local intrinsic dimensionality as an outlier measure reaches a better peak performance than other methods on several datasets and with several quality measures. Just as examples, we show the performance of all methods over all choices of the neighborhood size (k) on two datasets in terms of average precision and of ROC AUC (the measures are discussed by Campos et al. [10]) in Figs. 3 and 4.

On Parkinson (a smaller dataset), LID reaches optimal average precision (Fig. 3(a)) and ROC AUC (Fig. 3(b)) for a range of 6 choices of the neighborhood size and is still much better than all other methods on a larger range around these optimal values. The second example InternetAds is a medium-size (in this variant 1630 objects, 32 outliers), but high dimensional (1555 attributes) dataset. Here the quality of LID increases slightly (w.r.t. average precision, Fig. 4(a)) or remains rather stable (w.r.t. ROC AUC, Fig. 4(b)) when the neighborhood size approaches $k = 100$, in both measures outperforming the other methods by a large margin.

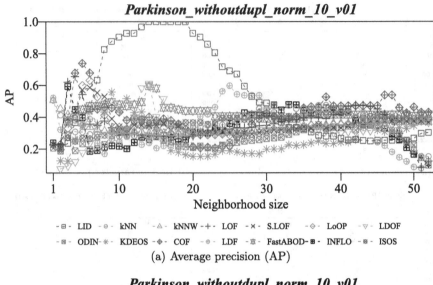

(a) Average precision (AP)

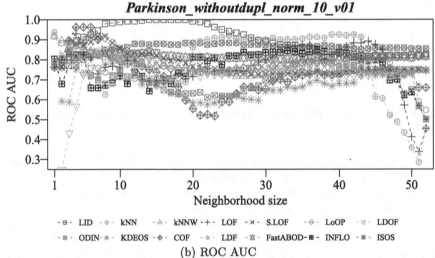

(b) ROC AUC

Fig. 3. Example: performance of the methods over all parameter settings w.r.t. to average precision and ROC AUC, dataset Parkinson (without duplicates, normalized, 10% outliers, random version 1).

However, individual results on individual datasets are a proof of concept at best. From the average performance over many datasets as depicted in Fig. 2 we can conclude more generally that the local intrinsic dimensionality as a measure of outlierness is in rough correlation with various established methods, while contributing different aspects to the corpus of outlier measures. Note that we kept the neighborhood size in the range that was tested by Campos et al. [10]

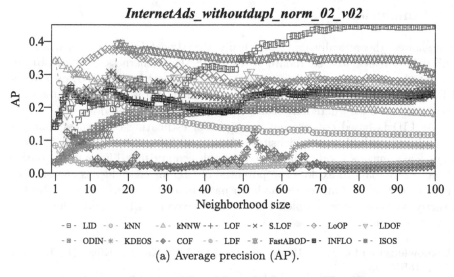

(a) Average precision (AP).

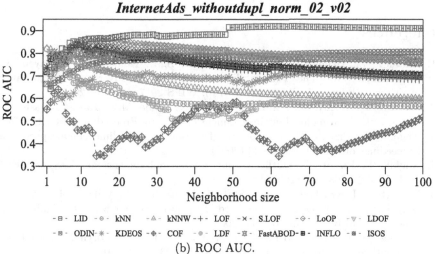

(b) ROC AUC.

Fig. 4. Example: performance of the methods over all parameter settings w.r.t. to average precision and ROC AUC, dataset InternetAds (without duplicates, normalized, 2% outliers, random version 2).

(up to $k = 100$). For most methods, the performance varies considerably with k, while there are no rules known on how to choose k. For LID (but also ISOS) however, we can expect that it performs better with larger k, in particular with $k \gg 100$ for larger datasets. In the interest of maintaining a comparable setting to Campos et al. [10], however, we kept k in the same range for all methods.

5   Conclusion

We showed theoretically and empirically that Local Intrinsic Dimensionality (LID) is a reasonable candidate to estimate the outlierness of data objects. Despite being outperformed by some of the other methods tested, on average LID is nevertheless of comparable quality for many datasets. At the same time, there are datasets, where LID clearly outperforms other methods.

As LID is based on different principles of estimating outlierness than the established methods (using density estimates in different variants), LID would most likely make a good addition to ensembles comprising different base outlier detectors [39,63]. To explore the potential of LID for ensembles is therefore an interesting topic for future research. In particular, it remains to be established exactly whether more complex formulations involving LID could achieve even better performance.

Acknowledgments. M. E. Houle supported by JSPS Kakenhi Kiban (B) Research Grant 18H03296.

References

1. Amsaleg, L., Bailey, J., Barbe, D., Erfani, S.M., Houle, M.E., Nguyen, V., Radovanović, M.: The vulnerability of learning to adversarial perturbation increases with intrinsic dimensionality. In: WIFS 2017, pp. 1–6 (2017)
2. Amsaleg, L., Chelly, O., Furon, T., Girard, S., Houle, M.E., Kawarabayashi, K., Nett, M.: Estimating local intrinsic dimensionality. In: Proceedings of KDD (2015)
3. Angiulli, F., Pizzuti, C.: Fast outlier detection in high dimensional spaces. In: Proceedings of PKDD, pp. 15–26 (2002)
4. Angiulli, F., Pizzuti, C.: Outlier mining in large high-dimensional data sets. IEEE TKDE **17**(2), 203–215 (2005)
5. Barnett, V., Lewis, T.: Outliers in Statistical Data, 3rd edn. Wiley, Hoboken (1994)
6. Beygelzimer, A., Kakade, S., Langford, J.: Cover trees for nearest neighbors. In: Proceedings of ICML, pp. 97–104 (2006)
7. Breunig, M.M., Kriegel, H.P., Ng, R., Sander, J.: LOF: identifying density-based local outliers. In: Proceedings of SIGMOD, pp. 93–104 (2000)
8. Camastra, F., Vinciarelli, A.: Estimating the intrinsic dimension of data with a fractal-based method. IEEE TPAMI **24**(10), 1404–1407 (2002)
9. Campello, R.J.G.B., Moulavi, D., Zimek, A., Sander, J.: Hierarchical density estimates for data clustering, visualization, and outlier detection. ACM TKDD **10**(1), 5:1–5:51 (2015)
10. Campos, G.O., Zimek, A., Sander, J., Campello, R.J.G.B., Micenková, B., Schubert, E., Assent, I., Houle, M.E.: On the evaluation of unsupervised outlier detection: measures, datasets, and an empirical study. Data Min. Knowl. Discov. **30**, 891–927 (2016)
11. Casanova, G., Englmeier, E., Houle, M., Kroeger, P., Nett, M., Schubert, E., Zimek, A.: Dimensional testing for reverse k-nearest neighbor search. PVLDB **10**(7), 769–780 (2017)
12. Costa, J.A., Hero, A.O.: Entropic graphs for manifold learning. In: 37th Asilomar Conference on Signals, Systems, and Computers, vol. 1, pp. 316–320 (2003)

13. de Vries, T., Chawla, S., Houle, M.E.: Density-preserving projections for large-scale local anomaly detection. KAIS **32**(1), 25–52 (2012)
14. Falconer, K.: Fractal Geometry: Mathematical Foundations and Applications. Wiley, Hoboken (2003)
15. Fraga Alves, M., de Haan, L., Lin, T.: Estimation of the parameter controlling the speed of convergence in extreme value theory. Math. Methods Stat. **12**(2), 155–176 (2003)
16. Grassberger, P., Procaccia, I.: Characterization of strange attractors. Phys. Rev. Lett. **50**, 346–349 (1983)
17. Grubbs, F.E.: Procedures for detecting outlying observations in samples. Technometrics **11**(1), 1–21 (1969)
18. Gupta, A., Krauthgamer, R., Lee, J.R.: Bounded geometries, fractals, and low-distortion embeddings. In: Proceedings of FOCS, pp. 534–543 (2003)
19. Hautamäki, V., Kärkkäinen, I., Fränti, P.: Outlier detection using k-nearest neighbor graph. In: Proceedings of ICPR, pp. 430–433 (2004)
20. Hawkins, D.: Identification of Outliers. Chapman and Hall, Boca Raton (1980)
21. Hein, M., Audibert, J.Y.: Intrinsic dimensionality estimation of submanifolds in R^d. In: Proceedings of ICML, pp. 289–296 (2005)
22. Hill, B.M.: A simple general approach to inference about the tail of a distribution. Ann. Stat. **3**(5), 1163–1174 (1975)
23. Houle, M.E.: Dimensionality, discriminability, density and distance distributions. In: Proceedings of ICDM Workshops, pp. 468–473 (2013)
24. Houle, M.E.: Local intrinsic dimensionality I: an extreme-value-theoretic foundation for similarity applications. In: Beecks, C., Borutta, F., Kröger, P., Seidl, T. (eds.) SISAP 2017. LNCS, vol. 10609, pp. 64–79. Springer, Cham (2017). https://doi.org/10.1007/978-3-319-68474-1_5
25. Houle, M.E.: Local intrinsic dimensionality II: multivariate analysis and distributional support. In: Beecks, C., Borutta, F., Kröger, P., Seidl, T. (eds.) SISAP 2017. LNCS, vol. 10609, pp. 80–95. Springer, Cham (2017). https://doi.org/10.1007/978-3-319-68474-1_6
26. Houle, M.E., Kashima, H., Nett, M.: Generalized expansion dimension. In: ICDM Workshop PTDM, pp. 587–594 (2012)
27. Houle, M.E., Ma, X., Nett, M., Oria, V.: Dimensional testing for multi-step similarity search. In: Proceedings of ICDM, pp. 299–308 (2012)
28. Houle, M.E., Ma, X., Oria, V.: Effective and efficient algorithms for flexible aggregate similarity search in high dimensional spaces. IEEE TKDE **27**(12), 3258–3273 (2015)
29. Houle, M.E., Ma, X., Oria, V., Sun, J.: Efficient algorithms for similarity search in axis-aligned subspaces. In: Traina, A.J.M., Traina, C., Cordeiro, R.L.F. (eds.) SISAP 2014. Lecture Notes in Computer Science, vol. 8821, pp. 1–12. Springer, Cham (2014). https://doi.org/10.1007/978-3-319-11988-5_1
30. Houle, M.E., Ma, X., Oria, V., Sun, J.: Query expansion for content-based similarity search using local and global features. ACM Trans. Multimed. Comput. Commun. Appl. (TOMM) **13**(3), 1–23 (2017)
31. Houle, M.E., Oria, V., Wali, A.M.: Improving k-nn graph accuracy using local intrinsic dimensionality. In: Beecks, C., Borutta, F., Kröger, P., Seidl, T. (eds.) SISAP 2017. LNCS, vol. 10609, pp. 110–124. Springer, Cham (2017). https://doi.org/10.1007/978-3-319-68474-1_8
32. Houle, M.E., Nett, M.: Rank-based similarity search: reducing the dimensional dependence. IEEE TPAMI **37**(1), 136–150 (2015)

33. Huisman, R., Koedijk, K.G., Kool, C.J.M., Palm, F.: Tail-index estimates in small samples. J. Bus. Econ. Stat. **19**(2), 208–216 (2001)
34. Jin, W., Tung, A.K.H., Han, J., Wang, W.: Ranking outliers using symmetric neighborhood relationship. In: Ng, W.-K., Kitsuregawa, M., Li, J., Chang, K. (eds.) PAKDD 2006. LNCS (LNAI), vol. 3918, pp. 577–593. Springer, Heidelberg (2006). https://doi.org/10.1007/11731139_68
35. Jolliffe, I.T.: Principal Component Analysis, 2nd edn. Springer, New York (2002). https://doi.org/10.1007/b98835
36. Karger, D.R., Ruhl, M.: Finding nearest neighbors in growth-restricted metrics. In: Proceedings of STOC, pp. 741–750 (2002)
37. Knorr, E.M., Ng, R.T.: Algorithms for mining distance-based outliers in large datasets. In: Proceedings of VLDB, pp. 392–403 (1998)
38. Kriegel, H.P., Kröger, P., Schubert, E., Zimek, A.: LoOP: local outlier probabilities. In: Proceedings of CIKM, pp. 1649–1652 (2009)
39. Kriegel, H.P., Kröger, P., Schubert, E., Zimek, A.: Interpreting and unifying outlier scores. In: Proceedings of SDM, pp. 13–24 (2011)
40. Kriegel, H.P., Schubert, M., Zimek, A.: Angle-based outlier detection in high-dimensional data. In: Proceedings of KDD, pp. 444–452 (2008)
41. Larrañaga, P., Lozano, J.A.: Estimation of Distribution Algorithms: A New Tool for Evolutionary Computation, vol. 2. Springer, New York (2002). https://doi.org/10.1007/978-1-4615-1539-5
42. Latecki, L.J., Lazarevic, A., Pokrajac, D.: Outlier Detection with Kernel Density Functions. In: Perner, P. (ed.) MLDM 2007. LNCS (LNAI), vol. 4571, pp. 61–75. Springer, Heidelberg (2007). https://doi.org/10.1007/978-3-540-73499-4_6
43. Levina, E., Bickel, P.J.: Maximum likelihood estimation of intrinsic dimension. In: Proceedings of NIPS, pp. 777–784 (2004)
44. Ma, X., Li, B., Wang, Y., Erfani, S.M., Wijewickrema, S.N.R., Schoenebeck, G., Song, D., Houle, M.E., Bailey, J.: Characterizing adversarial subspaces using local intrinsic dimensionality, pp. 1–15 (2018)
45. Ma, X., Wang, Y., Houle, M.E., Zhou, S., Erfani, S.M., Xia, S., Wijewickrema, S.N.R., Bailey, J.: Dimensionality-driven learning with noisy labels, pp. 1–10 (2018)
46. Navarro, G., Paredes, R., Reyes, N., Bustos, C.: An empirical evaluation of intrinsic dimension estimators. Inf. Syst. **64**, 206–218 (2017)
47. Papadimitriou, S., Kitagawa, H., Gibbons, P.B., Faloutsos, C.: LOCI: fast outlier detection using the local correlation integral. In: Proceedings of ICDE, pp. 315–326 (2003)
48. Pei, Y., Zaïane, O., Gao, Y.: An efficient reference-based approach to outlier detection in large datasets. In: Proceedings of ICDM, pp. 478–487 (2006)
49. Radovanović, M., Nanopoulos, A., Ivanović, M.: Reverse nearest neighbors in unsupervised distance-based outlier detection. IEEE TKDE **27**, 1369–1382 (2015)
50. Ramaswamy, S., Rastogi, R., Shim, K.: Efficient algorithms for mining outliers from large data sets. In: Proceedings of SIGMOD, pp. 427–438 (2000)
51. Romano, S., Chelly, O., Nguyen, V., Bailey, J., Houle, M.E.: Measuring dependency via intrinsic dimensionality, pp. 1207–1212 (2016)
52. Rousseeuw, P.J., Hubert, M.: Robust statistics for outlier detection. WIREs DMKD **1**(1), 73–79 (2011)
53. Rozza, A., Lombardi, G., Ceruti, C., Casiraghi, E., Campadelli, P.: Novel high intrinsic dimensionality estimators. Mach. Learn. **89**(1–2), 37–65 (2012)
54. Schubert, E., Gertz, M.: Intrinsic t-stochastic neighbor embedding for visualization and outlier detection. In: Proceedings of SISAP, pp. 188–203 (2017)

55. Schubert, E., Koos, A., Emrich, T., Züfle, A., Schmid, K.A., Zimek, A.: A framework for clustering uncertain data. PVLDB **8**(12), 1976–1979 (2015)
56. Schubert, E., Zimek, A., Kriegel, H.P.: Generalized outlier detection with flexible kernel density estimates. In: Proceedings of SDM, pp. 542–550 (2014)
57. Schubert, E., Zimek, A., Kriegel, H.P.: Local outlier detection reconsidered: a generalized view on locality with applications to spatial, video, and network outlier detection. Data Min. Knowl. Discov. **28**(1), 190–237 (2014)
58. Schubert, E., Zimek, A., Kriegel, H.-P.: Fast and scalable outlier detection with approximate nearest neighbor ensembles. In: Renz, M., Shahabi, C., Zhou, X., Cheema, M.A. (eds.) DASFAA 2015. LNCS, vol. 9050, pp. 19–36. Springer, Cham (2015). https://doi.org/10.1007/978-3-319-18123-3_2
59. Takens, F.: On the numerical determination of the dimension of an attractor. In: Braaksma, B.L.J., Broer, H.W., Takens, F. (eds.) Dynamical Systems and Bifurcations. LNM, vol. 1125, pp. 99–106. Springer, Heidelberg (1985). https://doi.org/10.1007/BFb0075637
60. Tang, J., Chen, Z., Fu, A.W., Cheung, D.W.: Enhancing effectiveness of outlier detections for low density patterns. In: Chen, M.-S., Yu, P.S., Liu, B. (eds.) PAKDD 2002. LNCS (LNAI), vol. 2336, pp. 535–548. Springer, Heidelberg (2002). https://doi.org/10.1007/3-540-47887-6_53
61. Wang, Y., Parthasarathy, S., Tatikonda, S.: Locality sensitive outlier detection: a ranking driven approach. In: Proceedings of ICDE, pp. 410–421 (2011)
62. Zhang, K., Hutter, M., Jin, H.: A new local distance-based outlier detection approach for scattered real-world data. In: Theeramunkong, T., Kijsirikul, B., Cercone, N., Ho, T.-B. (eds.) PAKDD 2009. LNCS (LNAI), vol. 5476, pp. 813–822. Springer, Heidelberg (2009). https://doi.org/10.1007/978-3-642-01307-2_84
63. Zimek, A., Campello, R.J.G.B., Sander, J.: Ensembles for unsupervised outlier detection: challenges and research questions. SIGKDD Explor. **15**(1), 11–22 (2013)

Graphs and Applications

Intrinsic Degree: An Estimator of the Local Growth Rate in Graphs

Lorenzo von Ritter[1,2(✉)], Michael E. Houle[1], and Stephan Günnemann[2]

[1] National Institute of Informatics,
2-1-2 Hitotsubashi, Chiyoda-ku, Tokyo 101-8430, Japan
meh@nii.ac.jp
[2] Technical University of Munich, Arcisstraße 21, 80333 Munich, Germany
lorenzo.ritter@tum.de, guennemann@in.tum.de

Abstract. The neighborhood size of a query node in a graph often grows exponentially with the distance to the node, making a neighborhood search prohibitively expensive even for small distances. Estimating the growth rate of the neighborhood size is therefore an important task in order to determine an appropriate distance for which the number of traversed nodes during the search will be feasible. In this work, we present the intrinsic degree model, which captures the growth rate of exponential functions through the analysis of the infinitesimal vicinity of the origin. We further derive an estimator which allows to apply the intrinsic degree model to graphs. In particular, we can locally estimate the growth rate of the neighborhood size by observing the close neighborhood of some query points in a graph. We evaluate the performance of the estimator through experiments on both artificial and real networks.

Keywords: Intrinsic dimensionality · Graph · Degree · Estimation

1 Introduction

Many real world data sets, such as the world wide web, social networks or the power grid, are structured as graphs [4]. Analyzing such data structures is therefore a relevant research topic. One important analysis is to estimate the growth rate of graphs. The neighborhood size of a node often grows exponentially with the search radius, making a neighborhood search prohibitively expensive even for small radii. Estimating the neighborhood size in order to decide on a feasible radius for the neighborhood search is therefore an important application.

In this paper, we present a model to calculate the growth rate of graphs that is based in the recent work in calculating the Local Intrinsic Dimensionality (LID) of data [10,17,18], a continuous expansion-based model of intrinsic dimensionality. LID and its precursor, the Generalized Expansion Dimension (GED) [11] have been effectively applied to similarity search problems, in particular in the use of dimensional testing for the early termination of search [7,12–15]. In the area of data mining and machine learning, their recent applications include the

© Springer Nature Switzerland AG 2018
S. Marchand-Maillet et al. (Eds.): SISAP 2018, LNCS 11223, pp. 195–208, 2018.
https://doi.org/10.1007/978-3-030-02224-2_15

following: guiding the local sparsification of features concurrently with similarity graph construction [16]; the analysis of non-functional dependencies among data features [24]; a theoretical analysis of the effect of adversarial perturbation on classification [2] and a practical deep neural network detection method for adversarial examples [21]; and a deep neural network classification strategy that monitors the dimensionality of subspaces during training and adapts the loss function so as to limit overfitting [22].

In this paper, we show that the equivalent of the local intrinsic dimensionality of high dimensional data is the local intrinsic degree of graphs, by extending the definition of LID from [17]. In Sect. 2, we give a brief overview of the LID model. In Sect. 3, we transform this model to the domain of graphs, show that it can fully characterize exponential growth functions and present an estimator. We evaluate the estimator through experiments on artificial and real data in Sects. 4 and 5, respectively. Finally, Sect. 6 provides a conclusion and an outlook for future work on this topic.

2   Intrinsic Dimensionality

The complexity of analyzing data sets often grows exponentially with the dimensionality of the data. However, some high-dimensional data sets can be represented in a lower-dimensional space with negligible loss of information. This lower-dimensional representation without loss of expressiveness is called the intrinsic dimensionality of the data, as opposed to the representational dimensionality, which is any higher-dimensional representation of the data.

In order to allow efficient computations on high-dimensional data sets, it is desirable to transform data to their intrinsic dimensionality. A crucial step for dimensionality reduction is to find or estimate the intrinsic dimensionality. Then, dimensionality reduction techniques can be applied to transform the data.

Recently, Houle [17,18] introduced the theoretical foundation of extracting the intrinsic dimensionality of data from a local perspective. The main idea is that the intrinsic dimensionality is related to the growth rate of the amount data points within an increasing proportion of a uniformly distributed data set. Therefore, it can be estimated by observing distances to neighbors from a specific point in the data distribution.

Importantly, this approach is based on analyzing the underlying distribution of the data instead of a specific data set. Moreover, since the analysis is performed locally, only the distribution of distances to nearby points is relevant for the estimation. Hence, no knowledge of the global distribution of the data is required.

The intrinsic dimensionality is defined as the normalized rate of increase of the neighborhood size, i.e.

$$\mathrm{ID}_F(x) \triangleq \lim_{\epsilon \to 0} \frac{F((1+\epsilon)x) - F(x)}{\epsilon \cdot F(x)} = \frac{x \cdot F'(x)}{F(x)} \tag{1}$$

where $F(x)$ is the distribution of the neighborhood size within distance x of a starting point. $\mathrm{ID}_F^* \triangleq \lim_{x \to 0} \mathrm{ID}_F(x)$ is called the local intrinsic dimensionality and describes the intrinsic dimensionality at the query point.

3 Intrinsic Degree

3.1 Transformation from the Intrinsic Dimensionality

The above formulation yields the intrinsic dimensionality for a polynomially growing neighborhood size of a query point. In graphs, however, the neighborhood size often grows exponentially. In this case, the local intrinsic dimensionality tends towards zero, as illustrated in the following example for the neighborhood size $G(y) = a^y$ at distance y.

$$\text{ID}_G^* = \lim_{y \to 0} \text{ID}_G(y) = \lim_{y \to 0} \frac{y \cdot \frac{d}{dy} a^y}{a^y} = \lim_{y \to 0} \frac{y \cdot a^y \ln a}{a^y} = \lim_{y \to 0} y \cdot \ln a = 0 \quad (2)$$

We can observe that for exponential neighborhood size functions, the intrinsic dimensionality does not yield any useful information. We have to find an alternative way to calculate the growth rate. Intuitively, the growth rate of a graph should be its average degree or equivalently the base a of the exponential neighborhood size function G.

In order determine this quantity, we find a relation between the exponential growth function $G(y)$ and a polynomial growth function $F(x)$. In particular, we assume that $G(y)$ was created by applying a logarithmic transformation, e.g. the natural logarithm, to the distance variable x of $F(x)$. This yields the relations

$$y \triangleq \ln(x) \qquad G(y) = G(\ln(x)) \triangleq F(x) \quad (3)$$

For estimating the intrinsic dimensionality of an exponential growth function $G(y) = a^y$, we could now transform the function $G(y)$ to the polynomial function $F(x) = a^{\ln x}$ according to the above relations. Subsequently, Eq. 1 can be used to calculate the intrinsic dimensionality.

$$\text{ID}_F(x) = \frac{x \cdot F'(x)}{F(x)} = \frac{x \cdot \frac{d}{dx} a^{\ln x}}{a^{\ln x}} = \frac{x \cdot \frac{d}{dy} a^{\ln x} \cdot \frac{dy}{dx}}{a^{\ln x}} = \frac{x \cdot a^{\ln x} \cdot \ln a \cdot \frac{1}{x}}{a^{\ln x}} = \ln a \quad (4)$$

Since the distance variable x dropped out of the above equation, we can also state that $\text{ID}_F^* = \ln a$.

Instead of transforming the exponential function and then computing the intrinsic dimensionality, we can also set up a formulation of the intrinsic dimensionality in the exponential regime.

$$\text{ID}_F(x) \triangleq \frac{x \cdot F'(x)}{F(x)} = \frac{x \cdot G'(\ln(x)) \cdot \frac{1}{x}}{G(\ln(x))} = \frac{G'(y)}{G(y)} \triangleq \ln(\text{IB}_G(y)) \triangleq \log\text{IB}_G(y) \quad (5)$$

Following our previous intuition that the growth rate of a graph should be its average degree or equivalently the base of its growth function, we call the counterpart of the intrinsic dimensionality in the exponential regime the intrinsic degree or intrinsic base interchangeably. We use the symbol IB in order to avoid mistaking it with the intrinsic dimensionality ID. As a short form for the logarithm of the intrinsic degree, we use logIB.

Evaluating the local intrinsic degree yields

$$\mathrm{logIB}_G^* = \lim_{y \to \ln(0)} \mathrm{logIB}_G(y) = \lim_{y \to -\infty} \frac{\frac{d}{dy} a^y}{a^y} = \lim_{y \to -\infty} \frac{a^y \ln a}{a^y} = \ln a. \qquad (6)$$

We observe that the result is the same as in Eq. 4, i.e. $\mathrm{ID}_F^* = \mathrm{logIB}_G^*$. In other words, transforming an exponential function to the polynomial regime and using the intrinsic dimensionality formulation yields the same result as using the intrinsic degree formulation and applying it directly to the exponential function.

Finally, we can transform the logarithm of the local intrinsic degree back to its original representation.

$$\exp(\mathrm{logIB}_G^*) = \mathrm{IB}_G^* = a \qquad (7)$$

As expected, the calculation shows that the local intrinsic degree and thus the growth rate of an exponential function is its base.

For the remainder of this paper, we are only interested in the local intrinsic base IB_G^*. Therefore, we omit to explicitly call it "local".

3.2 IB-Based Representation of Smooth Exponential Functions

Following the argument from [17], we can use the intrinsic degree IB_G to fully characterize an exponential function G under certain conditions.

Theorem 1 (Local IB Representation). *Let $G : \mathbb{R} \to \mathbb{R}$ be a real-valued function, and let $v \in \mathbb{R}$ be a value for which $\mathrm{IB}_G(v)$ exists. Let x and r be values for which x/r and $G(x)/G(r)$ are both positive. If G is non-zero and continuously differentiable everywhere in the interval $[\min\{x,r\}, \max\{x,r\}]$, then*

$$\frac{G(x)}{G(r)} = \mathrm{IB}_G(v)^{x-r} \cdot H_{G,v,w}(x), \quad \text{where}$$

$$H_{G,v,w}(x) \triangleq \exp\left(\int_x^r (\mathrm{logIB}_G(v) - \mathrm{logIB}_G(t))\, dt \right),$$

whenever the integral exists.

Proof. For any x and r for which x/r and $G(x)/G(r)$ are both positive,

$$G(x) = G(r) \cdot \exp\left(\ln\left(\frac{G(x)}{G(r)} \right) \right)$$

$$= G(r) \cdot \exp\left(\mathrm{logIB}_G(v) \cdot (x - r) - \mathrm{logIB}_G(v) \cdot (x - r) - \ln\left(\frac{G(r)}{G(x)} \right) \right)$$

$$= G(r) \cdot \mathrm{IB}_G(v)^{x-r} \cdot \exp\left(\mathrm{logIB}_G(v) \cdot \int_x^r dt - \int_x^r \frac{G'(t)}{G(t)} dt \right)$$

$$= G(r) \cdot \mathrm{IB}_G(v)^{x-r} \cdot \exp\left(\int_x^r \mathrm{logIB}_G(v) dt - \int_x^r \mathrm{logIB}_G(t) dt \right).$$

In the last line, we used $\mathrm{logIB}_G(x) = \frac{G'(x)}{G(x)}$ from Eq. 5. The integrals can be combined, which proves the statement. □

When the radius r tends towards the reference value v and x does not lie too far from r, the factor $H_{G,v,w}(x)$ vanishes. Therefore, Theorem 1 allows to characterize an exponential function G in the vicinity of a reference value v though its intrinsic degree $\mathrm{IB}_G(v)$. Further, when $v = 0$, the function G is fully characterized by its local intrinsic degree IB_G^*. The derivation and proofs are analogous to the representation of polynomial functions by their intrinsic dimensionality from [17] and therefore are omitted here.

3.3 Estimation of the Intrinsic Degree

We determine an estimator for the intrinsic base using maximum likelihood estimation, following the derivation of an estimator for the intrinsic dimensionality from [3]. For a query node in a graph, we assume that the distances to the k nearest neighbor nodes $y_1, \ldots, y_k$ are independently drawn from the random variable Y with support $[0, r)$. This corresponds to the use case where a query node of a graph is selected and the distances for all k nodes within radius r are recorded.

We want to find the most likely log intrinsic base $\log\mathrm{IB}_G^*$ that generated the distribution Y. The log-likelihood is given by

$$\mathcal{L}(\log\mathrm{IB}_G^* | y_1, \ldots, y_k) = \ln \prod_{i=1}^{k} G'(y_i | \log\mathrm{IB}_G^*) = \sum_{i=1}^{k} \ln G'(y_i | \log\mathrm{IB}_G^*), \quad (8)$$

where $G'(y)$ is the derivative of the cumulative distance distribution $G(y)$ and its log local intrinsic base is $\log\mathrm{IB}_G^*$. We determine $G'(y)$ by

$$G'(y) = \frac{dG(y)}{dy} = \frac{dF(x)}{dx}\frac{dx}{dy} = \frac{F(e^r)}{e^r} \cdot \mathrm{ID}_F^* \left(\frac{x}{e^r}\right)^{\mathrm{ID}_F^* - 1} \cdot \frac{de^y}{dy}$$

$$= \frac{G(r)}{e^r} \cdot \log\mathrm{IB}_G^* \left(\frac{e^y}{e^r}\right)^{\log\mathrm{IB}_G^* - 1} \cdot e^y = G(r) \cdot \log\mathrm{IB}_G^* \left(\frac{e^y}{e^r}\right)^{\log\mathrm{IB}_G^*}. \quad (9)$$

We used $F(x)/F(e^r) = (x/e^r)^{\mathrm{ID}_F^*}$ from [17] as well as the transformations from the previous section, i.e. $F(x) = G(y)$, $F(e^r) = G(r)$, and $\mathrm{ID}_F^* = \log\mathrm{IB}_G^*$.

We can now write the log-likelihood as

$$\mathcal{L}(\log\mathrm{IB}_G^* | y_1, \ldots, y_k) = k \cdot \ln G(r) + k \cdot \ln \log\mathrm{IB}_G^* + \log\mathrm{IB}_G^* \cdot \sum_{i=1}^{k} \ln \frac{e^{y_i}}{e^r}, \quad (10)$$

and its derivative with respect to $\log\mathrm{IB}_G^*$ as

$$\frac{\partial}{\partial \log\mathrm{IB}_G^*} \mathcal{L}(\log\mathrm{IB}_G^* | y_1, \ldots, y_k) = \frac{k}{\log\mathrm{IB}_G^*} + \sum_{i=1}^{k} (y_i - r). \quad (11)$$

We set the derivative of the log-likelihood to zero to get the maximum likelihood estimate of the logarithm of IB_G^*.

$$\widehat{\log\mathrm{IB}}_G^* = \underset{\log\mathrm{IB}_G^*}{\mathrm{argmax}}\, \mathcal{L}(\log\mathrm{IB}_G^*|y_1,\ldots,y_k) = \left(r - \frac{1}{k}\sum_{i=1}^{k} y_i\right)^{-1}. \tag{12}$$

Finally, we can calculate the maximum likelihood estimate of the intrinsic degree as $\widehat{\mathrm{IB}}_G^* = \exp(\widehat{\log\mathrm{IB}}_G^*)$. The complexity of the estimate is $\mathcal{O}(k)$.

4   Experiments on Artificial Networks

In order to evaluate the intrinsic degree estimator, we first test is on several artificially simulated graphs. In particular, we use the Erdős-Rényi model, the preferential attachment model with the Barabási-Albert algorithm and the Kallenberg-exchangeable graphon. As a ground truth, we will use the average degree of the generated graphs, which approximately gives the base of the neighborhood size growth function. All experiments were run using Matlab on an Ubuntu machine with an Intel Core i7-5930K CPU and 64 GB memory.

4.1   Erdős-Rényi Random Graphs

Introduction. The simplest graph to analyze is an undirected random graph $G(n,p)$ specified by the number of nodes n and the edge probability p, which was introduced by Gilbert, Erdős and Rényi [8,9]. In order to sample such a graph, we connect each node pair with probability p. For n large enough, the average degree converges to $n \cdot p$.

Experimental Setup. For the average degrees 1, 3, 5 and 7, we sample graphs with 10, 50, 100, 500, 1000, 5000, 10000 and 50000 nodes,[1] which was the largest possible size that our computational resources could handle. For each parameter combination, we sample one graph. For each graph, we randomly select 5 initial nodes, estimate the intrinsic degree with Eq. 12 and average over the estimates.

The estimator depends on the number of neighboring nodes that are included in the calculation, i.e. the k nearest neighbors. These can be implicitly given by the radius r up to which neighboring nodes are included. We use the radii 1 through 4 for our calculations.

Results and Discussion. Figure 1 contains the results of our experiments. We can observe that the estimator converges towards the average degree as we increase the number of nodes. While the convergence is better for smaller average degrees, it seems likely that, even for larger degrees, we will see a convergence if we further increase the number of nodes in the graph.

---

[1] Implementation from [6]: http://www.stats.ox.ac.uk/~caron/code/bnpgraph/.

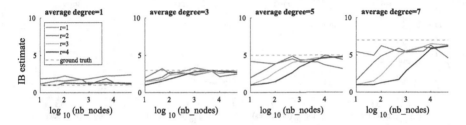

Fig. 1. IB estimates for random graphs of various average degrees. For each degree, we calculate the estimate for neighborhood radii 1 through 4.

Interestingly, using a larger neighborhood radius and therefore including more neighboring nodes in our estimate seems to have an adverse effect on the estimator. However, including only direct neighbors (i.e. using a radius of 1) does not always lead to a converging estimate, especially for large average degrees. The optimal choice seems to be to use neighbors up to distance 2 or 3 for the estimate.

4.2 Scale-Free Graphs

Introduction. As opposed to random graphs, the degree distribution of scale-free graphs follows a power law. Many real world graphs, such as social networks, power grids or the world wide web, exhibit such a degree distribution which makes the analysis of scale-free graphs an important research area [4].

A popular method to generate such graphs is the Barabási-Albert (BA) algorithm [1], which is based on the preferential attachment model [23]. This model generates a network by adding nodes one by one. Each new node connects to m existing nodes, while preferably connecting to nodes that have a large degree. This approach follows the "rich get richer" idea, i.e. nodes with a high degree are more likely to be connected to even more new nodes as the graph grows. Thus, scale-free graphs can be generated. Since m edges are added for each node, the average degree of the generated graph is $2 \cdot m$.

Experimental Setup. We use a Matlab implementation of the BA algorithm to generate scale-free graphs.[2] We use the values 1 through 4 for m and thus generate graphs with average degrees 2, 4, 6 and 8, respectively. For each m, we generate graphs containing 500, 1000, 5000, 10000 nodes. The Matlab implementation is very slow for graphs with more than 10000 nodes, which made larger experiments impossible due to time constraints.

For each m and each graph size, we sample 10 graphs. For each graph, we randomly select 10 initial nodes and calculate the intrinsic base estimate for neighborhood radii 1 through 5.

---

[2] https://de.mathworks.com/matlabcentral/fileexchange/11947-b-a-scale-free-network-generation-and-visualization.

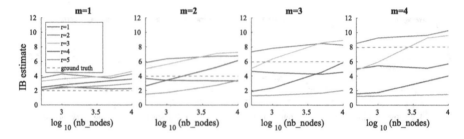

Fig. 2. The estimates for the intrinsic base for scale-free graphs generated by the BA algorithm with m 1 through 4.

Results and Discussion. The plots in Fig. 2 show the results of our experiments. We can observe that the estimator strongly depends on the radius r. For $r = 1$, the estimate is constant with respect to the number of nodes and slightly underestimates the average degree. When increasing the radius, the estimate first becomes larger and overestimates the average degree and then smaller to underestimate it. At the same time, the estimate shows the tendency to increase with the number of nodes. Finally, for radii greater or equal than 5, the estimates remains at a constant low value, particularly for large average degrees.

The intuition behind this phenomenon is that scale-free graphs have hubs, i.e. nodes with a very large degree. These hubs might be directly connected to most other nodes in the graph. Therefore, it is probable to reach such a hub with the first few steps from any initial node. When continuing the neighborhood search from the hub, the neighborhood size explodes. Therefore, the intrinsic base estimator will overestimate the average degree.

Interestingly, when further increasing the radius, the number of new neighborhood nodes starts to decrease. This can be explained through the "small world phenomenon" in scale-free graphs [25]. It can be experimentally shown that the average shortest path between any connected node pair in a scale-free graph is low even for very large networks. This means that a neighborhood search of a small radius will suffice to find nearly all nodes in the graph. Therefore, further increasing the radius does not significantly increase the number of new nodes, which in our case leads to a smaller intrinsic degree estimate.

In general, scale-free graphs introduce new difficulties to the estimation of the intrinsic degree. Using large radii yields bad results and is computationally expensive because most of the graph is visited during the neighborhood search. An appropriate trade-off might be to use the estimates for small radii and average over their results to obtain a final estimate. We will analyze this approach in more detail in the following section.

4.3   Graphons

Introduction. Graphons (an abbreviation for graph function) are graph distributions and are therefore of particular interest for evaluating the intrinsic degree. More specifically, they could allow to calculate the exact intrinsic degree according to Eq. 5 and compare our estimator to it. However, this would require knowing the neighborhood size function G as well as its derivative. Finding this function seems to be a combinatorial problem of exponential complexity and approximating it with simpler functions, e.g. through a random walk, proved to be too inaccurate. Therefore, we continue to use the average degree as a ground truth and leave the exact calculation of the intrinsic degree of a graphon for future work.

A graphon is a model for generating unweighted undirected graphs. It is defined as a function $W: [0,1]^2 \rightarrow [0,1]$, which maps every point on the unit square to a probability value. A finite unweighted graph with n nodes can be generated by sampling n values from the uniform distribution, i.e. $(u_i)_{i=1}^{n} \sim \mathcal{U}(0,1)$. An edge between the nodes i and j is then added with probability $W(u_i, u_j)$ [20].

The graphon model is very powerful due to its abstraction from a specific graph to a graph generating function. However, it can be proven that any graphon that follows the above definition will generate dense graphs [20]. Since many real world graphs are sparse, there is an interest in extending the graphon model to sparse graphs.

One such extension is the Kallenberg representation model, which was recently introduced by Caron and Fox [6]. The core of this approach is the node weight vector w, which assigns a weight (or sociability factor) to every potential node. The values of w are drawn from a Poisson point process, e.g. a generalized gamma process. The sum of all values in w determines the number of edges D^* that will be sampled for the graph, in particular $D^* \sim \text{Exp}(\sum_i w_i)$. Each edge is placed between two nodes based on their node weights w_i. This results in graph $G = (V, E)$, which consists of the sampled edges E and the all nodes that have at least one adjacent edge V.

The parameters of the Poisson point process have an important impact on the distribution of the node weight vector w and thus on the characteristics of graphs that can be generated. The authors of the model show that with certain parameters, sparse graphs will be generated where the degree distribution of the nodes follows a power law.

Experimental Setup. For our experiments, we use the generalized gamma process to generate the node weight vector, which depends on the three parameters α, σ and τ. We use the following parameters to create graph functions.

- $\alpha = 20, 40, 60, \ldots, 500$
- $\sigma = 0.1, 0.3, 0.5, 0.8$
- $\tau = 5, 10, 30, 50, 80, 100$

Each of the 600 parameter combinations defines a graph function. For each function, we draw 10 graph samples (see footnote 1) and on each of these, we choose 10 random initial nodes to estimate the intrinsic base.

Here, we use radii 1 trough 4 to evaluate the intrinsic base. Using a larger radius was not possible since some graphs had a high average degree and a larger neighborhood radius would include too many neighboring nodes, which exceeded our computational limitations.

As opposed to the previous models, we cannot explicitly specify the average degree or the number of nodes of the graphs that we sample. Instead, the average degree and the number of nodes depend on the choice of the parameters α, σ and τ. In order to overcome this limitation, we binned the generated graphs into groups by their average degree and used the centers of the average degree intervals as a ground truth.

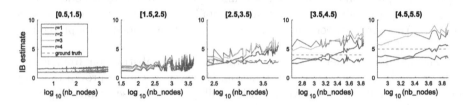

Fig. 3. Estimation of the intrinsic base for graphon samples of the average degree intervals $[0.5, 1.5)$, $[1.5, 2.5)$, $[2.5, 3.5)$, $[3.5, 4.5)$ and $[4.5, 5.5)$.

Results and Discussion. An observation of the experiment results in Fig. 3 shows that when only including neighbors up to radius 1 (i.e. direct neighbors), the average degree is underestimated by the intrinsic base estimator, especially for large average degrees. For radii 2 and 3, the estimator overestimates the average degree instead. When using larger distances, the estimator does not remain constant but seems to increase with the number of nodes in the graph.

In order to further improve our estimation, we remember that we are estimating the local intrinsic degree, i.e. the growth rate of the neighborhood size at an infinitesimal distance from the query node. However, in order to allow any estimation, we also have to include neighboring nodes that are further away. Nevertheless, in order to focus on the neighborhood growth in close proximity to the query node, it might be beneficial to give more weight to nodes that are closer to it.

This idea can easily be implemented by averaging over the two estimations for radii of 1 and 2. Since the estimate for radius 2 also includes all nodes within radius 1 of the origin, averaging over both estimates is equivalent to giving twice the weight to the estimates for neighbors within radius 1.

The results are plotted in Fig. 4, which shows that the average of the estimations for radii 1 and 2 gives a good approximation of the true average degree. It is remarkable that we can give such an accurate estimation of the average degree by just observing neighbors of up to distance 2.

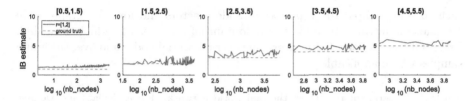

Fig. 4. Averaging over the estimates of radii 1 and 2 gives a good approximation.

5 Experiments on Real World Networks

Data Sets. After having tested our estimator on synthetic data, we perform experiments with several real world graphs. We use the following data sets of unweighted and undirected graphs.

- Yeast [5]: A yeast protein interaction network. Each node in the graph is a protein of the budding yeast. If there is a known interaction to another protein, these two are connected by an edge.[3]
- Enron [19]: An email connection network from the Federal Energy Regulatory Commission. Nodes in the graph are email addresses, which are connected by an undirected edge if at least one email was sent between two email addresses.[4]
- IMDB [4]: An actor collaboration graph generated from a bipartite movie-actor graph. Actors are represented as nodes and connected to other actors if the starred together in at least one movie.[5]

Figure 5 visualizes the degree distributions of the three data sets. Table 1 contains the sizes and average degrees of the graphs.

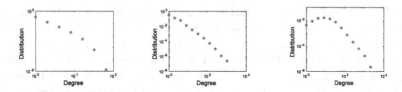

Fig. 5. The degree distributions for the yeast (left), Enron (center) and IMDB (right) graphs.

Experimental Setup. For each data set, we sample 10 random initial nodes and estimate the intrinsic base for radii 1, 2 and 3. Since the size of the IMDB data set exceeds our computational limitations, we create smaller subsets of the graph, using 10000, 50000 and 100000 randomly selected nodes (i.e. actors). For

---

[3] https://www.cise.ufl.edu/research/sparse/matrices/Pajek/yeast.html.
[4] https://snap.stanford.edu/data/email-Enron.html.
[5] https://www.cise.ufl.edu/research/sparse/matrices/Pajek/IMDB.html.

each size, we sample 10 graphs with random actors and for each graph again estimate the intrinsic base for 10 random initial nodes using radii 1 through 3. The results are reported after averaging over the initial nodes and over the graph samples where applicable.

Table 1. Experiment results on the real world networks. For each data set, we give the estimated intrinsic degree for radii 1, 2 and 3, as well as the respective normalized mean absolute errors (nMAE).

	Number of nodes	Average degree	Estimated IB			nMAE		
			r = 1	r = 2	r = 3	r = 1	r = 2	r = 3
yeast	2361	6.08	4.61	5.94	6.52	0.2425	**0.0235**	0.0716
enron	36692	10.02	2.99	10.55	9.24	0.7014	**0.0525**	0.0783
imdb_10k	10000	3.43	3.05	4.21	4.56	**0.1114**	0.2269	0.3307
imdb_50k	50000	9.14	4.45	8.99	9.72	0.5119	**0.0163**	0.0636
imdb_100k	100000	16.15	4.98	10.36	10.86	0.6915	0.3590	**0.3279**
imdb	896308	129.33	–	–	–	–	–	–

Results and Discussion. The experiment results can be found in Table 1. For each data set, we list the size and the average degree, as well as the estimated intrinsic degree for radii 1 through 3. Additionally, for each radius we calculate the normalized mean absolute error (nMAE), i.e. the mean absolute error of the prediction divided by the average degree.

We can observe that the best estimates are obtained for neighborhood radii larger than 1. Estimates on the IMDB data set generally seem to be worse than for the other two graphs. One reason might be the different degree distributions (see Fig. 5). While the yeast and Enron graphs follow a power law degree distribution, the degree of the IMDB data is more evenly distributed. It seems possible that power law-distributed graphs yield better results because most nodes have similar small degrees and only very few nodes have large degrees. Nevertheless, during the experiments on Erdős-Rényi random graphs, we saw that the estimator also performs well on evenly distributed networks. A more detailed analysis on the impact of graph characteristics on the performance of the estimator is left open for future work.

Averaging the estimates of various radii as done during the experiments on artificial graphs did not significantly improve the accuracy.

6   Conclusion and Future Work

In this work, we laid the theoretical groundwork for locally estimating the growth rate of graphs through their intrinsic degree. We transferred the intrinsic dimensionality model to exponential functions, which yielded the intrinsic degree.

Further, we showed that the intrinsic degree fully characterizes an exponential growth function and introduced an estimator using maximum likelihood estimation. Extensive experiments on both artificial and real world networks showed that the estimator can capture the approximate growth rate of graphs but its accuracy varies depending on the graph structure. We found that averaging over the predictions for various radii can improve the estimations.

Further work is required to better understand the estimator. A detailed analysis on the influence of graph characteristics on the performance of the estimator could help to predict for which types of graphs the estimator will yield good results. Such work could include analyzing weighted and directed graphs.

Moreover, using weighted averages of the estimates for different radii in order to improve the predictions seems a promising approach. Investigating the influence of the characteristics of a graph on the radius weights could contribute to making more reliable estimates.

Finally, in this paper we use the average degree as a ground truth for the intrinsic degree estimate. For graphons, a better choice would be the true intrinsic degree according Eq. 5, which requires finding a feasible solution or approximation of the neighborhood size function G as well as its derivative.

Acknowledgments. This research was supported in part by the Technical University of Munich - Institute for Advanced Study, funded by the German Excellence Initiative and the European Union Seventh Framework Programme under grant agreement no 291763, co-funded by the European Union. M. E. Houle was supported by JSPS Kakenhi Kiban (B) Research Grant 18H03296.

References

1. Albert, R., Barabási, A.L.: Topology of evolving networks: local events and universality. Phys. Rev. Lett. **85**(24), 5234–5237 (2000)
2. Amsaleg, L., Bailey, J., Barbe, D., Erfani, S.M., Houle, M.E., Nguyen, V., Radovanović, M.: The vulnerability of learning to adversarial perturbation increases with intrinsic dimensionality. In: WIFS 2017, pp. 1–6 (2017)
3. Amsaleg, L., Chelly, O., Furon, T., Girard, S., Houle, M.E., Kawarabayashi, K., Nett, M.: Estimating local intrinsic dimensionality. In: SIGKDD, pp. 29–38 (2015)
4. Barabási, A.L., Albert, R.: Emergence of scaling in random networks. Science **286**(5439), 509–512 (1999)
5. Bu, D.: Topological structure analysis of the protein-protein interaction network in budding yeast. Nucl. Acids Res. **31**(9), 2443–2450 (2003)
6. Caron, F., Fox, E.B.: Sparse graphs using exchangeable random measures. J. Roy. Stat. Soc.: Ser. B (Stat. Methodol.) **79**(5), 1295–1366 (2017)
7. Casanova, G., Englmeier, E., Houle, M., Kroeger, P., Nett, M., Schubert, E., Zimek, A.: Dimensional testing for reverse k-nearest neighbor search. PVLDB **10**(7), 769–780 (2017)
8. Erdős, P., Rényi, A.: On random graphs I. Publicationes Mathematicae Debrecen **6**, 290 (1959)
9. Gilbert, E.N.: Random graphs. Ann. Math. Stat. **30**(4), 1141–1144 (1959)
10. Houle, M.E.: Dimensionality, discriminability, density and distance distributions. In: ICDMW, pp. 468–473 (2013)

11. Houle, M.E., Kashima, H., Nett, M.: Generalized expansion dimension. In: ICDMW, pp. 587–594 (2012)
12. Houle, M.E., Ma, X., Nett, M., Oria, V.: Dimensional testing for multi-step similarity search. In: ICDM, pp. 299–308 (2012)
13. Houle, M.E., Ma, X., Oria, V.: Effective and efficient algorithms for flexible aggregate similarity search in high dimensional spaces. TKDE **27**(12), 3258–3273 (2015)
14. Houle, M.E., Ma, X., Oria, V., Sun, J.: Efficient algorithms for similarity search in axis-aligned subspaces. In: Traina, A.J.M., Traina, C., Cordeiro, R.L.F. (eds.) SISAP 2014. LNCS, vol. 8821, pp. 1–12. Springer, Cham (2014). https://doi.org/10.1007/978-3-319-11988-5_1
15. Houle, M.E., Ma, X., Oria, V., Sun, J.: Query expansion for content-based similarity search using local and global features. TOMM **13**(3), 25:1–25:23 (2017)
16. Houle, M.E., Oria, V., Wali, A.M.: Improving *k*-nn graph accuracy using local intrinsic dimensionality. In: Beecks, C., Borutta, F., Kröger, P., Seidl, T. (eds.) SISAP 2017. LNCS, vol. 10609, pp. 110–124. Springer, Cham (2017). https://doi.org/10.1007/978-3-319-68474-1_8
17. Houle, M.E.: Local intrinsic dimensionality I: an extreme-value-theoretic foundation for similarity applications. In: Beecks, C., Borutta, F., Kröger, P., Seidl, T. (eds.) SISAP 2017. LNCS, vol. 10609, pp. 64–79. Springer, Cham (2017). https://doi.org/10.1007/978-3-319-68474-1_5
18. Houle, M.E.: Local intrinsic dimensionality II: multivariate analysis and distributional support. In: Beecks, C., Borutta, F., Kröger, P. (eds.) SISAP 2017. LNCS, vol. 10609, pp. 80–95. Springer, Cham (2017). https://doi.org/10.1007/978-3-319-68474-1_6
19. Klimt, B., Yang, Y.: The Enron corpus: a new dataset for email classification research. In: Boulicaut, J.-F., Esposito, F., Giannotti, F., Pedreschi, D. (eds.) ECML 2004. LNCS (LNAI), vol. 3201, pp. 217–226. Springer, Heidelberg (2004). https://doi.org/10.1007/978-3-540-30115-8_22
20. Lovász, L.: Large Networks and Graph Limits, Colloquium Publications, vol. 60. American Mathematical Society (2012)
21. Ma, X., Li, B., Wang, Y., Erfani, S.M., Wijewickrema, S.N.R., Schoenebeck, G., Song, D., Houle, M.E., Bailey, J.: Characterizing adversarial subspaces using local intrinsic dimensionality. In: ICLR, pp. 1–15 (2018)
22. Ma, X., Wang, Y., Houle, M.E., Zhou, S., Erfani, S.M., Xia, S., Wijewickrema, S.N.R., Bailey, J.: Dimensionality-driven learning with noisy labels. In: ICML, pp. 1–10 (2018)
23. Price, D.D.S.: A general theory of bibliometric and other cumulative advantage processes. J. Am. Soc. Inf. Sci. **27**(5), 292–306 (1976)
24. Romano, S., Chelly, O., Nguyen, V., Bailey, J., Houle, M.E.: Measuring dependency via intrinsic dimensionality. In: ICPR, pp. 1207–1212 (2016)
25. Travers, J., Milgram, S.: An experimental study of the small world problem. Sociometry **32**(4), 425 (1969)

Advanced Analytics of Large Connected Data Based on Similarity Modeling

Tomáš Skopal[1], Ladislav Peška[1(✉)], Irena Holubová[1],
Petr Paščenko[2], and Jan Hučín[2]

[1] Department of Software Engineering, Faculty of Mathematics and Physics,
Charles University, Prague, Czech Republic
{skopal,peska,holubova}@ksi.mff.cuni.cz
[2] Profinit EU, s.r.o., Prague, Czech Republic
{petr.pascenko,jan.hucin}@profinit.eu

Abstract. Collecting various types of data about users/clients in order
to improve the services and competitiveness of companies has a long
history. However, these approaches are often based on classical statistical methods and an assumption of limited computational power. In this
paper we introduce the vision of our applied research project targeting to
the financial sector. Our main goal is to develop an automated software
solution for similarity modeling over big and semi-structured graph data
representing behavior of bank clients. The main aim of similarity models is to improve the decision process in risk management, marketing,
security and related areas.

Keywords: Similarity modeling · Big Data · Analysis of graph data
Transactional data · Linked data

1 Introduction

The success of leading on-line enterprises like Google, Facebook or Amazon arises
from the perfect knowledge of their clients. For this purpose, huge amount of
information (e.g., users' click-streams) is being collected. Evaluation of these
large data allows to establish advanced models of customer preferences, client-
to-product relationships, client-to-client relationships etc. and enables making
optimal business decisions and achieving high profits.

Similar approaches can be applied also in other areas, such as, e.g., financial
sector or telecommunications. Banks, insurance companies, and telecommunication operators all posses large user-based data collections that are, however,
analyzed and studied mainly using classical statistical methods and based on
the assumptions of limited computational power and lack of data. Nonetheless, similarity and relationship modeling has the potential to disclose valuable,
yet unknown information and, hence, fundamentally improve existing business
decision processes, user experience, risk management, detection of fraudulent
behavior and more. At the same time, novel Big Data processing technologies

S. Marchand-Maillet et al. (Eds.): SISAP 2018, LNCS 11223, pp. 209–216, 2018.
https://doi.org/10.1007/978-3-030-02224-2_16

provide capability to process significantly larger volumes of data in a far greater detail than the most powerful relational databases.

The main goal of our applied research project is the development of an automated software solution intended for similarity modeling over big semi-structured network data. In the first phase, we focus on the financial sector, namely data describing financial transactions in a bank. We aim to create a complex view of a client involving his/her environment, historical and current behavior, social relationships etc. In the next phase the complex client description will be utilized in various pattern mining scenarios to support bank's decision processes. The aim of this paper is to introduce the current approaches in the industry and related open problems (see Sect. 2), and the proposal of our solution together with the related research results to be utilized/extended in our approach (see Sect. 3).

2   Industry Best Practices and Challenges

The wide range of data mining tasks in banks, insurance and telecommunication enterprises are common classification problems. In such cases, the task is to predict probability of a certain behavior (e.g., churn or falling to default) or to detect predefined events (e.g., incoming salary or fraud). Therefore, classical statistical models like logistic regression, naive Bayes classifier, tree-based learning or neural networks can be applied for such tasks. However, input features commonly used in these models are often very simple, mostly based on client's demographics or simple aggregations of past behavior (e.g., gender, age, risk class, average income etc.). And the quality of the learned models is, naturally, bounded by the quality of input features.

Although there are many options to construct more complex and more relevant features for particular tasks (e.g., relations between clients, processing series of client's transaction, clustering clients by their respective areas of interest etc.), such features turn out to be too challenging for routine processing due to their complexity and excessive volume. Furthermore, classical statistical models sometimes provide only weak or moderate predictors due to the flat nature of the input data. For example, considering social relationships between clients, one cannot simply use the whole social network as a reasonable input feature vector of a model. Likewise, estimating of behavior can be done more efficiently and accurately by finding similar patterns in transactions, user actions, text etc.

2.1   Advanced Analytics

In the state of the art data modeling tools used in the finance industry the main focus is given to the predictive modeling problem. It consists of three components: the features, the model and the target. Let us assume that we want to predict whether a particular client will repay his/her loan or (s)he eventually falls into default. We use the features (client age, income, loan amount etc.) and fit a model (logistic regression, decision tree, neural network, etc.) based on historical data (previous loans and their defaults) in order to predict future behavior of the client. This process can be improved in three ways: (1) we can develop better models, (2) we can develop better features, and (3) we can make the process of modeling more user convenient.

In the last three decades, enormous scientific effort was given to develop highly advanced predictive models (such as deep neural networks, support vector machines, complex decision trees, ensembles etc.). With a decent delay, they are implemented for commercial use in convenient tools (like, e.g., SAS[1], IBM SPSS[2], RapidMiner[3], and many others). With rather greater delay, they are absorbed by industry practitioners who often prefer model transparency and simplicity to its novelty. Though, we do not see much space for improvements in this field.

On the contrary, the features and analytics have not changed much. Most of the prediction features are either of simple categorization (e.g., age, sex, address etc.) and/or straightforward database aggregations are used (e.g., previous loans count, average income, number of years in the bank etc.). In general, due to conceptual simplicity and computational efficiency of SQL based data warehouses, the features are calculated as aggregations of transactional databases.

With the Big Data technologies, advances in the field of feature engineering can be expected for two reasons: (1) Big Data clusters provide significantly higher computational power to mine more advanced features and (2) unlike the SQL, Big Data technologies operates on full scale programming language (Java, Scala, Python), so more advanced techniques, such as similarity search, can be easily implemented.

We believe that analytics based on similarity modeling on clients (or other entities) together with building a social network is a challenge and could represent a substantial step forward in the development of software applications related to data science (at least in the financial sector).

2.2   Limitations of SQL Servers

Suppose we want to discover household relations (similarities) of bank clients based on the transactional data. It turns out, that simple direct transaction from person A to person B is present for less than half of actual household members. On the other hand, the events of two consecutive payments on the same location in a short time window are relatively frequent for family members (imagine family mall visit). When we want to aggregate the count of such events (A and B both have a payment on geographically close places during the same time window) we need to perform a series of expensive joins of which the most complex is the self join in transaction table followed by user defined filtered aggregation (time windowing, close locality testing etc.). For a query like this we need both, big computation power as well as feasible programming language. Conventional SQL RDBMS can not be employed for a task of this sort.

---

[1] https://www.sas.com/en_us/home.html.

[2] https://www.ibm.com/products/spss-statistics.

[3] https://rapidminer.com/.

3   Proposed Similarity Framework

In our bank client similarity framework, we model a client using her/his profile as well as using transaction data that determine the clients actual financial behavior. Based on the behavior we believe the clients can be classified (or just clustered) into groups of clients with similar behavior. The typical motivation in the banking environment for such classification is the identification of risk (clients with good/bad prognosis of paying debts). However, in this research project we aim at neutral group modeling of clients with similar behavior that is unique/specific for the respective group.

3.1   Temporal Bipartite Graphs

While the profile data can be used to simple modeling of client similarity on profile attributes, the transaction data promise more sophisticated modeling. Transaction data is time-ordered sequence of financial transactions, such as credit card payments, ad-hoc or periodical bank transfers, ATM withdrawals etc. The transaction log can be interpreted as a temporal bipartite graph with client nodes on one side and service categories on the other (see Fig. 1). Existing approaches of transforming topological similarity into a latent feature vector [5,6] can be utilized and modified to incorporate temporal relevance as additional input. There is a wide range of possible models of temporal relevance, ranging from simple windowing or weighting schemes, to more complex temporal distance models or ensembles of detected *concept drift* points inducing heterogeneous edge types.

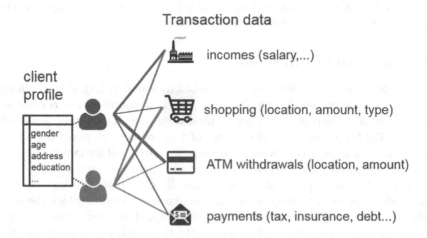

Fig. 1. Client data available to a bank displayed as (temporal) bipartite graph.

3.2   Multi-dimensional Time Series

Alternatively, the transaction log can be split into multiple time series per client, each time series identified by the client and a type of transaction (feature). In each feature series, we do not limit its elements to numeric values (amounts of money transferred) but there can be also category data generally characterizing the service related to the transaction (e.g., purchase in a luxury shoe store in the downtown). At the most detailed level, each client is thus represented by a multi-dimensional time series that can be directly used to measure similarity with another client either for separate feature-to-feature matching, or all-to-all features matching, including correlations between features. However, to achieve a robust descriptor, we assume quantization and/or aggregation of the input time series using windowing, such that similar behavior patterns are identified regardless of (noisy) details. See Fig. 2.

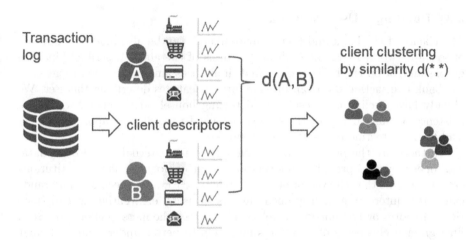

Fig. 2. Clients modeled by multi-dimensional time series.

Quantization. In the research of similarity search in time series datasets, there have been developed models that quantize real values to discrete values or even to symbols [2]. This allows to abstract from absolute numbers to more robust descriptors (like in other data engineering areas).

Windowing. Splitting the time series into blocks is another crucial task when modeling time series descriptors. We will consider traditional windowing, i.e., overlapping and non-overlapping windows, while the width and position of windows can be static or based on analysis of rapid change in the signal, similarly as cuts are detected in video scenes.

Transformation. The time series split into windows can be optionally transformed from the time domain into frequency domain (using Fourier or Wavelet transformation) or can be otherwise aggregated (e.g., to simple histogram). While the time dimension is partially preserved in the sequence of windows, the descriptors within windows can be more distinctive or robust after the transformation.

Similarity Functions. For variable-length time series there have been developed many distance function, while as the most versatile it is usually used the dynamic time warping distance (DTW) [4]. For the multi-dimensional case, the similarity scores on individual single-dimensional time series (for particular features) can be combined, or there can be utilized a distance function that matches the entire multi-dimensional time series [3]. This allows to cope with fine-grain correlations between different features (their time series, respectively).

3.3   Building a User Network

The clients of the bank and their "environment" can be also modeled as a sort of social network. It can be built on the basis of the above described bipartite graphs and/or on the basis of evaluated similarities amongst clients. In general, the bank transaction data can provide further features describing the user. We already have preliminary results on detecting household members, where the existence of mutual transactions plays a crucial role. Similar approaches may lead to the detection of co-workers, friends etc.

Furthermore the network can be extended with external linked open data, e.g., open registers providing information about companies, shops, institutions etc. We assume that the vector of user's preference towards specific retail brands can detect interesting finer-grained clusters of users, e.g., enthusiasts to (particular) sports or technology, travelers, high street shoppers, gamers etc. Such finer-grained clustering of customers may help to better understand particular decisions made by the members of the cluster.

In general, having such an enhanced network of users and their relations, we can base its analysis on re-use and extension of various existing approaches having related aims. The best known is probably *social selling*, i.e., establishing and utilizing social connections to sell products (such as, e.g., framework *HeteroSales* [7] which builds a social network of companies on the basis of LinkedIn data and combines it with sales records to find new potential customers). Another related area is, e.g., *viral marketing* dealing with finding a set of individuals to maximize the word-of-mouth propagation of a brand [8]. Or there are approaches which analyze the behavior of traders in financial markets in order to detect, e.g., abnormal trading events [9]. Complex analysis of the structure of social networks, such as *community detection* [10], can also bring interesting relevant information. Similarly, *node and edge prediction* in graphs [5,6] can help us in discovering "hidden" information, including labels, nodes and edges.

Another key task is to detect changes in the regular behavioral patterns. We may utilize, e.g., differences of window-based temporal node2vec vectors of a particular user, or modified RNN networks for video scene detection [1] to detect concept drifts in transactional behavior. Collaborative filtering can be applied subsequently to cluster the drifts. We expect to receive indicators like, e.g., moving, employment lost or change, marriage, child birth etc. Furthermore, time-based collaborative filtering on the descriptors of transaction series may further disclose some risky user behavior, i.e., users with similar contemporary behavior fall to default later on.

4   Conclusion

In this paper we have presented a vision of a software solution for similarity modeling over big and semi-structured graph data representing environment and behavior of bank clients. The proposed framework utilizes similarity model for bank clients based on their financial behavior (transactional data). The model aims at developing bank social network where similar clients are connected by edges. The client similarity supposed to include temporal bipartite graph model and/or multi-dimensional time series.

Acknowledgments. This work was supported in part by the Technology Agency of the Czech Republic (TAČR) project no. TH03010276 and by Czech Science Foundation (GAČR) project no. 17-22224S.

References

1. Byeon, W., Breuel, T.M., Raue, F., Liwicki, M.: Scene labeling with LSTM recurrent neural networks. In: 2015 IEEE Conference on Computer Vision and Pattern Recognition (CVPR), pp. 3547–3555, June 2015
2. Camerra, A., Shieh, J., Palpanas, T., Rakthanmanon, T., Keogh, E.: Beyond one billion time series: indexing and mining very large time series collections with iSAX2+. Knowl. Inf. Syst. **39**(1), 123–151 (2014)
3. Cao, D., Liu, J.: Research on dynamic time warping multivariate time series similarity matching based on shape feature and inclination angle. J. Cloud Comput. **5**(1), 11 (2016)
4. Ding, H., Trajcevski, G., Scheuermann, P., Wang, X., Keogh, E.: Querying and mining of time series data: experimental comparison of representations and distance measures. Proc. VLDB Endow. **1**(2), 1542–1552 (2008)
5. Dong, Y., Chawla, N.V., Swami, A.: Metapath2vec: scalable representation learning for heterogeneous networks. In: Proceedings of the 23rd ACM SIGKDD International Conference on Knowledge Discovery and Data Mining, KDD 2017, New York, NY, USA, pp. 135–144. ACM (2017)
6. Grover, A., Leskovec, J.: Node2vec: scalable feature learning for networks. In: Proceedings of the 22nd ACM SIGKDD International Conference on Knowledge Discovery and Data Mining, KDD 2016, New York, NY, USA, pp. 855–864. ACM (2016)

7. Hu, Q., Xie, S., Zhang, J., Zhu, Q., Guo, S., Yu, P.S.: Heterosales: utilizing hetero-
    geneous social networks to identify the next enterprise customer. In: Proceedings
    of the 25th International Conference on World Wide Web, WWW 2016, Republic
    and Canton of Geneva, Switzerland, pp. 41–50. International World Wide Web
    Conferences Steering Committee (2016)
8. Liu, Q., Xiang, B., Chen, E., Xiong, H., Tang, F., Yu, J.X.: Influence maximization
    over large-scale social networks: a bounded linear approach. In: Proceedings of the
    23rd ACM International Conference on Conference on Information and Knowledge
    Management, CIKM 2014, New York, NY, USA, pp. 171–180. ACM (2014)
9. Wang, J., Zhou, S., Guan, J.: Detecting potential collusive cliques in futures mar-
    kets based on trading behaviors from real data. Neurocomputing 92, 44–53 (2012)
10. Zhang, J., Cui, L., Yu, P.S., Lv, Y.: BL-ECD: broad learning based enterprise
    community detection via hierarchical structure fusion. In: Proceedings of the 2017
    ACM on Conference on Information and Knowledge Management, CIKM 2017,
    New York, NY, USA, pp. 859–868. ACM (2017)

Towards Similarity Models in Police Photo Lineup Assembling Tasks

Ladislav Peska[1(✉)] and Hana Trojanova[2]

[1] Faculty of Mathematics and Physics,
Charles University, Prague, Czech Republic
`peska@ksi.mff.cuni.cz`
[2] Faculty of Philosophy, Charles University, Prague, Czech Republic

Abstract. Photo lineups play a significant role in the eyewitness identification process. Lineups are used to provide evidence in the prosecution and subsequent conviction of suspects. Unfortunately, there are many cases where lineups have led to the incorrect identification and conviction of innocent suspects. One of the key factors affecting the incorrect identification is the lack of lineup fairness, i.e. that the suspect differs significantly from other candidates. Although the process of assembling a fair lineup is both highly important and time-consuming, only a handful of tools are available to simplify the task.

In this paper, we follow our previous work in this area and focus on defining and tuning the inter-person similarity metric that will serve as a base for a lineup candidate recommender system. This paper proposes an inter-person similarity metric based on DCNN descriptors of candidates' photos and their content-based features, which is further tuned by the feedback of domain experts. The recommending algorithm further considers the need for uniformity in lineups. The proposed method was evaluated in a realistic user study focused on lineup fairness over solutions proposed by domain experts.

Results shown indicate that the precision of the proposed method is similar to the solutions proposed by domain experts and therefore the approach may significantly reduce the amount of manual work needed for assembling photo lineups.

Keywords: Photo lineups · Recommender systems · Inter-person similarity

1 Introduction and Motivation

Evidence from eyewitnesses often plays a significant role in criminal proceedings. An important part of eyewitness identification is a police lineup (either lineup in natura, or a photo lineup, i.e., demonstration of objects' photographs). Lineups are assembled by placing a suspect within a group of persons known to be innocent (fillers) and are used to verify (or disprove) police suspicions of the suspects' guilt. The problem is that the suspect might be the culprit, but might also be innocent and in some cases, lineups played a role in the conviction of an innocent suspect [12].

There are many sources of error in eyewitness identifications. Some errors are caused by features of the recalled past event (e.g., distance from the scene, lighting conditions) and the context of the witness (e.g., level of attention or fear). Such features cannot be controlled in general. Controllable variables include the method of questioning, construction of the lineup, interaction with investigators, and more [13].

© Springer Nature Switzerland AG 2018
S. Marchand-Maillet et al. (Eds.): SISAP 2018, LNCS 11223, pp. 217–225, 2018.
https://doi.org/10.1007/978-3-030-02224-2_17

One of the principal recommendations for inhibiting errors in identification is to assemble lineups according to the lineup fairness principle [1, 9]. In general, fair lineups should ensure that the suspect is not substantially different from the fillers. All fillers should have similar visual attributes to the suspect to provide an appropriate level of uncertainty to unreliable eyewitnesses[1]. Eyewitnesses may be overconfident in their identification when fillers are too different, even in cases where an incorrect suspect is presented [3, 14]. Since the similarity metric is latent, lineup fairness is usually assessed on the basis of data obtained from *"mock witnesses"*, i.e., persons who have not seen the offender, but received a textual description of him/her. A lineup is considered fair (unbiased) if mock witnesses cannot reliably identify a suspect based only on a brief description.

1.1   Related Work

Assembling photo lineups, i.e., finding appropriate candidates to fill the lineup for a particular suspect is a challenging and time-consuming task involving the exploration of large databases of candidates. In recent years, some research projects [7, 15] as well as commerce activities[2] aimed to simplify the process of eyewitness identifications. They mostly focused on the lineup administration, but they did not support intelligent lineup assembling.

We approach the problem of lineup assembling from the recommender systems perspective, i.e., we aim to recommend suitable candidates for a particular suspect to the forensic technicians who assemble lineups. To the best of our knowledge, there were no previous attempts to utilize recommending principles in the lineup assembling task, except for our own previous work [11]. In this initial work, we summarized the lineup candidates' recommendation task and its specifics, proposed a dataset of lineup candidates, and evaluated two recommending methods based on the nearest neighbor model: one utilized the similarity of a person's content-based attributes such as age or appearance (CB), while the other method used the similarity of visual descriptors w.r.t. VGG-face DCNN network [10]. Both approaches were evaluated by domain experts selecting the best fitting candidates out of the recommended ones.

In this paper, we extend the previous work in several directions. We proposed a combination of both similarity metrics and tuned the final similarity to better explain the previous choices of domain experts. We also updated the recommendation process to incorporate the need for lineup uniformity. Finally, the method was evaluated in a user study with mock witnesses against the lineups proposed by domain experts.

The main contributions of this paper are:

- A proposed lineup candidates similarity metric and consistency enhancing recommending algorithm *LineRec*.
- An evaluation of *LineRec* w.r.t. lineup fairness over expert-created lineups.
- A dataset of candidate selections available for future off-line experiments.

---

[1] psychology.iresearchnet.com/forensic-psychology/eyewitness-memory/lineup-size-and-bias/.

[2] e.g., elineup.org.

2 Materials and Methods

2.1 Dataset of Lineup Candidates

Although there are several commercial lineup databases (see footnote 2), one should approach carefully while utilizing them due to the problem of localization (differences in ethnic groups, common clothing patterns, haircuts, make up trends etc.). As a general principle to inhibit the bias caused by strangers in the lineup, candidates' datasets should follow the same localization as the suspects. We base our lineups assembling research in the localization context of the Czech Republic. Although the majority of the population is Caucasian, mostly of Czech, Slovak, Polish, and German nationalities, there are large Vietnamese and Romany minorities which make lineup assembling more challenging.

In order to follow the correct localization, we collected the dataset of candidate persons from the wanted and missing persons application of the Police of the Czech Republic[3]. In total, we collected data about 4,423 missing or wanted males. All records contained a portrait photo, nationality, age, and appearance characteristics such as: (facial) hair color and style, eye color, figure shape, tattoos, and more. Detailed descriptions as well as dataset statistics can be found in our previous work [11].

We utilized the results of previous experiments to extend the dataset by explicit similarity relations among candidates as follows. In the previous experiments, domain experts were asked to select the best lineup candidates for a given suspect from a list of top-20 most similar candidates based either on visual or CB features. We consider a pair of persons as explicitly related if at least one expert put both of them into a single lineup and explicitly unrelated if candidates were seen, but not selected.

2.2 Models of Similarity

While defining similarity among suspects and lineup candidates, there are two principal similarity beds. First, more traditional sources of information are categorical features describing personal characteristics as mentioned above. CB features are included in the dataset, but may also be extracted from the description of offenders in real-world cases.

Other sources of information are visual descriptors of persons' photos. In this case, we utilized a pre-trained VGG-Face deep convolutional network for facial recognition problems [10]. VGG-Face was optimized on a multi-classification problem on the dataset of 2,622 celebrities with 1,000 images each. The network achieved state-of-the-art performance ratings on several face recognition benchmarks. We utilized the last fully connected layer of the network as a visual descriptor of each lineup candidate.

In the previous work [11], both sources of similarity were evaluated separately. Although visual-based similarity performed better than the one based on CB features, both methods provided some relevant results and therefore we aimed on merging them into a single similarity metric in this work. Furthermore, previous results also showed

---

[3] aplikace.policie.cz/patrani-osoby/Vyhledavani.aspx.

the need for fine-tuning the final similarity as the relevant candidates were selected almost uniformly across the list of top-20 candidates.

Based on the previous findings, we aimed to represent all persons by a vector of their latent features. Latent features were learned via a Siamese network with the input features consisted of both CB and visual descriptors and the optimization criterion was based on the knowledge of the relatedness of the pairs of candidates. The structure of the network is depicted on Fig. 1. The network was modelled and optimized in TensorFlow via Adam optimizer [5] with the following Siamese loss (1).

$$LOSS_{Siam} = \sum_{(o_i, o_j, y_{i,j}) \in T_s} \begin{cases} \text{IF } y_{i,j} = 1 : \ \|f_i - f_j\|_2^2 \\ \text{ELSE: } \max\left(C - \|f_i - f_j\|_2, 0\right)^2 \end{cases} \tag{1}$$

Where T_s denotes a train set, f_i and f_j are latent feature vectors of candidates o_i and o_j respectively, $y_{i,j}$ is an indicator, whether candidates o_i and o_j are related to each other, $\|\cdot\|_2$ is an L2-norm and C is a minimal margin constant. Roughly speaking, $LOSS_{Siam}$ penalizes any differences in related candidates' descriptors as well as the lack of sufficient difference among unrelated pairs.

The train set of candidate pairs T_s was constructed to contain both pairs explicitly denoted as related by experts (positive example), pairs considered to be similar w.r.t. CB or visual similarity, but unrelated according to experts opinion (negative example) and distant pairs of candidates w.r.t. both similarity metrics (negative example). As the explicit relatedness data were available only for a portion of candidates, we decided to utilize additional implicit positive examples created from visually highly similar pairs of candidates, which did not appear in the previous experiments and therefore, expert based evaluation was not available for them. We consider this to be an auxiliary criterion of relatedness as the explicit relatedness of visually similar candidates was quite high in general. We suppose that the importance of this auxiliary criterion should decrease over time, as more explicit relatedness feedback should be available from newly assembled lineups. As the final similarity metric of candidates c_i and c_j, we utilized a cosine similarity of their latent feature vectors (2).

$$s_{i,j} = \frac{f_i \cdot f_j}{\|f_i\|_2 \times \|f_i\|_2} \tag{2}$$

2.3   Recommending Lineup Candidates

Based on the survey among domain experts [11], we opted to incorporate the principle of lineup uniformity into the process of recommending lineup candidates, i.e., in addition to maximizing suspect to filler similarity (2), we aim to also maximize the total inter-person similarity of the lineup. This observation was rather surprising as it goes against the common practice of recommender systems, which generally aims to maximize diversity of the recommended list [6]. However, as the purpose of lineups is to increase the uncertainty of unreliable eyewitnesses, it is quite a reasonable requirement. The recommending algorithm (denoted as *LineRec* in the evaluation) is a

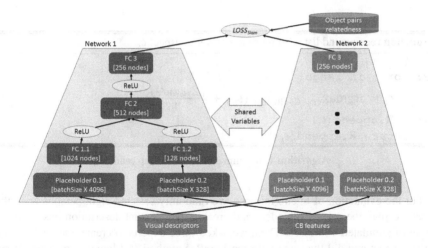

Fig. 1. Schema of the Siamese network for learning candidate's latent factors. The network contains three fully connected layers with ReLU activations. In the first layer, inputs from visual descriptors and CB features are treated separately (*FC1.1* and *FC1.2*) and merged in the second layer (*FC2*). The last fully connected layer (*FC3*) is considered as the vector of latent features of each candidate. The network is optimized w.r.t. Siamese loss (1) evaluated over both explicitly and implicitly related and unrelated pairs of candidates.

greedy weighted optimization of both criteria and is depicted on Algorithm 1. The algorithm is a variant of inversed Maximal Margin Relevance [2], with hyperparameter λ denoting relative importance of total inter-person similarity. During experiments, minimal margin C and inter-person similarity λ were held constant at 5.0 and 0.25 respectively.

3   Evaluation

In this paper, we aimed to evaluate fairness of the lineups produced by *LineRec* method over lineups created manually by the domain experts (forensic technicians). *LineRec* utilizes candidates' latent features to select closest matches to the suspect as well as already selected lineup fillers. A similar strategy is usually applied by domain experts while creating lineups manually (although the latent features resides in the expert's mind in this case), which is a common contemporary praxis. However, the manual creation of lineups is a tedious task[4]. Therefore, our primary research question is, up to which extent, such manual work can be replaced by automatic processing. Apart from technical details (i.e. assuring that selected candidates are clearly innocent), which so far cannot be automated, the main concern is the fairness of proposed lineups.

---

[4] During lineup assembling process, invited experts used a *LineIT* tool (available from herkules.ms. mff.cuni.cz/lineit) to browse candidates according to selected CB features. The average time to create a lineup using this tool was eleven minutes, however the task may become even more demanding in case of larger datasets or suspects with rare features.

Function recommend list of top-k candidates for suspect c_s

```
1:  R = []
2:  for i in range(k):
```
$$3: \quad c_i = \mathrm{argmax}_{\forall c_j \in C - (R \cup \{c_s\})} \left(s_{s,j} + \lambda \frac{\sum_{\forall c_r \in R} s_{r,j}}{|R|} \right)$$
```
4:      R.append(c_i)
5:  return R.append(c_s)
```

Algorithm 1: recommendation of lineup candidates.

For the evaluation of lineup fairness, we randomly selected eight persons from the dataset to play the role of offenders and created their textual description based on the output of simulated witnesses. Then, we asked three experts to create lineups for each of the suspect and did the same with the *LineRec* method. All lineups consisted of six persons in total, which is the most common in practice, so the process can be also seen as a top-five candidates' recommendation. Lineups constructed for the same suspect by different generators (*LineRec* method and Experts 1, 2 and 3) were quite different from each other with the average Jaccard similarity of 0.02. Finally, we conducted an on-line study asking volunteers to identify a suspect based on the description[5]. Each participant was asked to evaluate eight lineups, one for each suspect, while the selection of the lineup generator was random for each suspect.

There were in total 125 participants of the study, however, we removed those, who did not complete the whole study, or did not pay enough attention to descriptions (≤ 20 s per lineup) resulting into 108 participants. As for the demographics, all participants were Czech or Slovak with a small bias towards males (57%), reasonable age distribution and quite high bias towards higher education (46% with university degree).

3.1   Results

The total volume of suspects' identification was 864, out of which the true suspect was identified by participants in 487 cases. The overall per-user precision of lineups, i.e., the ratio of incorrectly identified suspects, was 0.44 ± 0.19. The overall per-generator precision was slightly higher for experts (0.435, 0.464 and 0.461 for Experts 1 to 3) than for *LinRec* method (0.419), however the differences were not statistically significant (p-value ≥ 0.304).

Surprisingly, we did not found any significant relation between age, gender, education or average time needed for decision and the correctness of the suspect identification either in general or per-generator. Nonetheless, the differences among lineups were quite distinctive. Table 1 depicts the precision score and effective lineup size (ES) [8] for each pair of lineup and generator. Effective lineup size is defined as follows:

---

[5] The evaluation GUI is available from herkules.ms.mff.cuni.cz/lineit-eval.

$$ES = |\mathbf{R}| - |\mathbf{R}\backslash\mathbf{R}_0| \sum_{c_i \in \mathbf{R}} \max\left(0, \frac{1}{|\mathbf{R}\backslash\mathbf{R}_0|} - f_i\right) \qquad (3)$$

Where $\mathbf{R}_0$ is the set of never selected candidates and f_i is the frequence of selecting candidate c_i. The intuition behind ES is to calculate, what should be the selection frequency of each candidate $(1/|\mathbf{R}\backslash\mathbf{R}_0|)$ and how much each candidate lags behind it. The sum of lags is then subtracted from the nominal lineup size $|\mathbf{R}|$. Therefore, in contrast to the precision score, which only calculates ratio of suspect to all other candidates, ES evaluates contribution of each candidate separately and is widely accepted by the forensic psychology community. Results illustrate that *LineRec* method scored well primarily in ES, where its results were best in one case and second best in 5 of 7 cases. Therefore, we may conclude that the performance of proposed method is fairly stable w.r.t. relative ES and usually provides a wide range of acceptable candidates.

Table 1. Results of precision and ES respectively. Best results are bold, second best underlined.

Lineup	Proposed method	Expert 1	Expert 2	Expert 3
1	0.34/2.70	**0.60/3.90**	0.36/2.45	0.11/1.32
2	<u>0.68/3.91</u>	**0.75/4.19**	0.33/2.67	0.50/3.50
3	<u>0.44/2.58</u>	0.26/2.05	0.55/1.91	**0.75/3.50**
4	**0.74/3.74**	0.19/1.76	<u>0.72/**4.11**</u>	0.59/2.91
5	0.05/1.16	0.13/1.38	<u>0.24/1.71</u>	**0.39/2.56**
6	0.30/2.48	**0.81/4.13**	0.33/2.33	<u>0.56/3.81</u>
7	<u>0.64/3.90</u>	0.53/3.76	0.76/**4.43**	**0.80**/3.75
8	0.25/**2.27**	0.23/1.69	**0.40**/2.20	0.12/1.48
AVG (ES)	2.84 ± 0.96	2.86 ± 1.24	2.73 ± 1.00	2.85 ± 0.99

4 Discussion and Conclusions

The primary goal of our work is to examine the usability of automated methods for lineup candidates' recommendation and utilize them in the process of lineups assembling. In this paper, we proposed a *LineRec* method aiming to learn a similarity metric based on visual and CB descriptors of candidates w.r.t. the explicit relatedness evaluation of experts and consider the lineup uniformity requirement in the recommendation process. *LineRec* method was evaluated w.r.t. lineup fairness over expert-created lineups representing common contemporary praxis. The evaluation demonstrated a similar average performance for both expert-created and *LineRec*-based lineups. However, there were rather high differences in the fairness of each lineup.

We further investigated the case of lineups 5 and 8, where none of the lineups provided reasonable fairness (ES ≤ 3.5). In both cases, two factors played an important role. First, suspects were members of an ethnical minority and therefore, fewer potentially relevant candidates were available. Second, in both cases, the textual description contained a highly selective feature (hedgehog haircut and problematic skin), which was hard to find in the limited dataset. Therefore, one direction of our future work is to extend the dataset of available candidates and let the administrator guide the similarity metric to focus more on a specific feature, e.g., via the similarity of image segments. Another option would be to apply a similar approach as in an artist style transfer [4] problem and alter other candidates' photos to better comply with the specific feature.

Thanks to the very small overlap of generated lineups and comparability of results, there is a high probability of a synergic effect in the case of cooperation of domain experts and *LineRec*-like methods. Therefore, we plan to extend the *LineIT* tool by such features and evaluate its effectiveness. Last but not least, due to the lack of test data in the similarity learning phase, we did not aim to experiment with different network structures or hyperparameter tuning and opted for rather simple solutions. However, as the test set of mock witness selections is available now, we plan to re-visit this problem and aim to propose improved similarity learning approaches w.r.t. off-line evaluations.

Acknowledgements. This work was supported by the Czech grants GAUK-232217 and GACR-17-22224S. Source codes and raw results are available from github.com/lpeska/LineRec.

References

1. Brigham, J.C.: Applied issues in the construction and expert assessment of photo lineups. Appl. Cogn. Psychol. **13**, S73–S92 (1999)
2. Carbonell, J., Goldstein, J.: The use of MMR, diversity-based reranking for reordering documents and producing summaries. In: 21st Annual International ACM SIGIR Conference on Research and Development in Information Retrieval, SIGIR 1998 (1998)
3. Charman, S.D., et al.: The dud effect: adding highly dissimilar fillers increases confidence in lineup identifications. Law Hum. Behav. **35**, 479–500 (2011)
4. Gatys, L.A., et al.: Image style transfer using convolutional neural networks. In: IEEE Conference on Computer Vision and Pattern Recognition (2016)
5. Kingma, D.P., Ba, J.L.: Adam: a method for stochastic optimization. In: 2015 International Conference on Learning Representations, pp. 1–15 (2015)
6. Kunaver, M., Požrl, T.: Diversity in recommender systems – a survey. Knowl.-Based Syst. **123**, 154–162 (2017)
7. MacLin, O.H., et al.: PC_Eyewitness and the sequential superiority effect: computer-based lineup administration. Law Hum. Behav. **29**, 303–321 (2005)
8. Malpass, R.S., Lindsay, R.C.: Measuring line-up fairness. Appl. Cogn. Psychol. **13**, S1–S7 (1999)
9. Mansour, J.K., et al.: Evaluating lineup fairness: variations across methods and measures. Law Hum. Behav. **41**, 103 (2017)
10. Parkhi, O.M., et al.: Deep face recognition. In: 2015 Proceedings of the British Machine Vision Conference (2015)
11. Peska, L., Trojanova, H.: Towards recommender systems for police photo lineup. In: ACM International Conference Proceeding Series (2017)
12. Smith, A.M., et al.: Fair lineups are better than biased lineups and showups, but not because they increase underlying discriminability. Law Hum. Behav. **41**, 127 (2017)

13. Steblay, N., et al.: Eyewitness accuracy rates in police showup and lineup presentations: a meta-analytic comparison. Law Hum. Behav. **27**, 523–540 (2003)
14. Tredoux, C.: A direct measure of facial similarity and its relation to human similarity perceptions. J. Exp. Psychol. Appl. **8**, 180 (2002)
15. Valentine, T.R., et al.: How can psychological science enhance the effectiveness of identification procedures? An international comparison. Public Interest Law Report (2007)

Privacy-Preserving String Edit Distance with Moves

Shunta Nakagawa[1], Tokio Sakamoto[2], Yoshimasa Takabatake[1], Tomohiro I[1], Kilho Shin[3], and Hiroshi Sakamoto[1(✉)]

[1] Kyushu Institute of Technology, 680-4 Kawazu, Iizuka, Fukuoka 820-8502, Japan
{s_nakagawa,takabatake,tomohiro,hiroshi}@donald.ai.kyutech.ac.jp
[2] ThomasLab Inc., 680-41 Kawazu, Iizuka, Fukuoka 820-0067, Japan
sakamoto@thomas-lab.jp
[3] University of Hyogo, 7-1-28 Minatojima-Minami, Kobe 650-0047, Japan
yshin@ai.u-hyogo.ac.jp

Abstract. We propose the first two-party protocol for securely computing an extended edit distance. The parties possessing their respective strings x and y want to securely compute the *edit distance with move operations* (EDM), that is, the minimum number of insertions, deletions, renaming of symbols, or substring moves required to transform x to y. Although computing the exact EDM is NP-hard, there exits an almost linear-time algorithm within the approximation ratio $O(\lg^* N \lg N)$ for $N = \max\{|x|, |y|\}$. We extend this algorithm to the privacy-preserving computation enlisting the homomorphic encryption scheme so that the party can obtain the approximate EDM without revealing their privacy under the *semi-honest model*.

1 Introduction

1.1 Motivation and Related Works

Edit distance is a well-established metric for measuring the dissimilarities between two strings. The rapid progress of sequencing technology has expanded the range of application of edit distance to personalized genomic medicine, disease diagnosis, and preventive treatment (for example, see [1]). A fundamental problem in these applications is represented by the edit-distance computation in a multi-party computation where parties possess their private strings. Lately, several efficient algorithms have been proposed for computing the edit distances in the communication models [5,8,15,24], which significantly reduce the communication overhead. A personal genome is, however, ultimate individual information that enables the identification of its owner, and thus the party cannot share personal genomic data as plaintext.

In literature, various privacy-preserving protocols based on cryptographic strategies have been proposed to quantify the similarities between two strings [2–4,9,14,20,29]. To evaluate a certain function f in a party, we can design a secure protocol based on Yao's *garbled circuit* (GC) [28] or on a homomorphic encryption (HE) system. GC enables two parties to securely compute $f(x,y)$ for any

© Springer Nature Switzerland AG 2018
S. Marchand-Maillet et al. (Eds.): SISAP 2018, LNCS 11223, pp. 226–240, 2018.
https://doi.org/10.1007/978-3-030-02224-2_18

probabilistic polynomial-time function f. Yao's original GC was not *reusable* (i.e., a new circuit needed to be generated for every single evaluation of $f(x, y)$ to prevent unauthorized disclosures); therefore, several reusable GCs have been proposed (for example, see [12]). However, it is difficult to directly use GC for our problem because $f()$ has variable arity, which depends on the input strings. Therefore, we employ another strategy: HE, which permits a certain operation over ciphertexts. The Paillier encryption system [18] enables us to perform additive operations for two encrypted integers without decryption. HE is suitable for developing an outsourced database in the multi-party model because of its semantically secure property, that is, any ciphertext $E(S)$ reveals nothing about the plaintext S even if an adversary can observe a number of pairs $(S_i, E(S_i))$ on his choice of S_i.

Using characteristics such as unlimited execution of a certain operation and the robustness for reuse, practical algorithms for securely computing the edit distance $d(x, y)$ and its variants have been proposed. Inan et al. [14] designed a symmetric three-party protocol for securely computing $d(x, y)$ where two-parties enlist the help of a reliable third party. Rane and Sun [20] improved the three-party protocol to an asymmetric two-party protocol, where one party has limited computational capacity and upload bandwidth whereas the other party is more powerful with better computational capacity and upload bandwidth. This constraint intends a situation in which a client possessing a single string x and a server possessing a number of strings y_i jointly compute $d(x, y_i)$. However, even now a variant of the privacy-preserving edit distance computation still remains an open question.

The extended metric is called the *edit distance with moves* (EDM) where any substring move in a unit cost is allowed in addition to the usual operations. Although computing the exact EDM is NP-hard [23], Cormode and Muthukrishnan [10] proposed an almost linear-time approximation algorithm for EDM. A string x is transformed into a characteristic vector v_x consisting of nonnegative integers referring to the frequency of particular substrings. The algorithm guarantees $L_1(v_x, v_y) = O(d_*(x, y) \lg^* N \lg N)$ for the exact EDM $d_*(x, y)$ and $N = \max\{|x|, |y|\}$. Because $\lg^* N$ hardly increases[1], we employ $L_1(v_x, v_y)$ as an approximate $d_*(x, y)$. Based on the EDM, we can approximately find a set of maximal common substrings between two strings, which is applicable for plagiarism detection. Consider two unambiguously similar strings $x = a^N b^N$ and $y = b^N a^N$ where one is transformed to the other by a single *cut-and-paste* operation. In contrast to the case of EDM, which gives us the preferable result $d_*(x, y) = 1$, the standard edit distance reports the undesirable result $d(x, y) = 2N$. The powerful similarity measure helps us to efficiently compress the original strings while preserving the capabilities of random access and the similarity search to the strings (for example, see [13,16,17,26]). Our proposal for securely computing EDM can widely extend the framework to ciphertexts.

---

[1] Here, $\lg^{(1)} N = \lg N$, $\lg^{(i+1)} N = \lg (\lg^{(i)} N)$ for $i \geq 1$, and $\lg^* N = \min\{i \mid \lg^{(i)} N \leq 1\}$. Thus, $\lg^* N \leq 5$ for $N \leq 2^{65536}$.

In practice, we can canonically assume that the two parties who own their private information are interested in maintaining the confidentiality of their secrets. However, we cannot assume that a third party keen on keeping their secrets; therefore, they can effectively work as a trusted third party.

1.2 Our Contribution

For a two-party secure computation of EDM, we need to design two secure protocols for computing a sharable naming function on substrings and the L_1-distance on integer vectors. The results that we obtained for each subproblem are described as follows.

Sharable Naming Function: On computing the approximate EDM, a string x is transformed to an ordered tree T_x identical to the derivation tree of a CFG in which any internal node u has a *consistent* name $f(u)$, that is, $f(u) = f(v)$ iff the sequence of the names of the children of u and v are equal. For a single string, such a naming function f is determined as follows: Let $[x]$ be the set of all substrings in x, and let π be a permutation of $[x]$. For any $z \in [x]$, we define $f(z) = k$ iff $\pi[k] = z$. Using this f, we can assign consistent names for all nodes in T_x. For two strings x and y, we can similarly construct f by a permutation π of $[x] \cup [y]$, which assigns consistent names for T_x and T_y. However, in our problem, the two parties should not reveal any substring of their private strings. Assuming a fully homomorphic encryption system (FHE) [11,12] that enables an arbitrary number of additions and multiplications over ciphertexts, we can design a secure protocol for f (see [9]). However, FHE is not practical because iterated multiplications impose a heavy overhead in terms of time and space complexity. Thus, we design a lightweight protocol for sharable f based on the Boneh-Goh-Nissim encryption system (BGN) [7] that enables an arbitrary number of additions and only one multiplication over ciphertexts. Our protocol significantly reduces the overhead as compared with several related techniques based on bit-wise encryption (e.g., Greater Than-Strong Conditional Oblivious Transfer (GT-SCOT [6]) and bit-decomposition [22,27]).

Lightweight L_1-Distance Computation: After computing T_x and T_y, they independently compute the characteristic vectors v_x and v_y such that $v_x[\ell]$ ($1 \leq \ell \leq n$) is the frequency of the name a_ℓ in T_x, where n is the number of different symbols appearing in T_x and T_y. Then, $L_1(v_x, v_y) = \sum_{\ell=1}^{n} |v_x[\ell] - v_y[\ell]|$ is a reasonable approximation of EDM [10]. This problem is reduced to securely computing the absolute value $E(|p-q|)$ from two encrypted non-negative integers $E(p)$ and $E(q)$ without decryption. Using earlier results [21], we can exactly and approximately solve this problem in a two-party protocol. In the exact method, p in the $\lg n$ bits is transformed to a bit-vector $B_p \in \{0,1\}^M$ for some $M >> n$ such that the first p bits of B_p are 1 and the others are 0. Clearly, $|p - q| = L_2(B_p, B_q)^2 = \sum_{i=1}^{M} (B_p[i]^2 + B_q[i]^2 - 2B_p[i]B_q[i])$. Note that, since either p or q is a plaintext, we can obtain $L_2(B_p, B_q)^2$ without decryption. However, the exact method is not practical due to the exponentially large communication overhead. In the approximate method, B_p is embedded into a smaller space;

however, we still cannot avoid constructing the explicit B_p [21]. Based on the additive property in the Paillier encryption system[2] [18], we propose a novel two-party protocol for computing $|p - q|$ in $O(1)$ round complexity and reasonable communication complexity.

In the following sections, we propose two protocols based on HE for the above problems. We also show the correctness, security, and complexity and give the comparison results. Finally, we mention a problem that remains unsolved.

2   Edit Distance with Moves

2.1   Definition

Let Σ be a finite set of alphabet symbols and Σ^* be its closure. The set of all strings of the length N is denoted by Σ^N. The length of a string S is denoted by $|S|$. For simplicity, the cardinality of a set U is also denoted by $|U|$. The i-th symbol of S is denoted by $S[i]$ for $i \in [1, |S|]$.

EDM $d_*(S, S')$ is the length of the shortest sequence of edit operations to transform S to S', where the permitted operations are insertion, deletion, and renaming of a symbol at any position and a substring move with unit cost. This distance measure is a *metric*. The distance between two strings is zero iff they are identical. Every operation has a unit-cost inverse, then $d_*(S, S') = d_*(S', S)$; and any distance defined by the above operations with unit-cost must obey the triangle inequality. Unfortunately, computing the exact EDM $d_*(S, S')$ is intractable even if renaming operations are not allowed. The following sections explain this further.

Theorem 1 ([23]). *Given the two strings S and S' and the integer $m \in \mathbb{N}$, using only three unit-cost operations of insertion, deletion, and substring move, it is NP-complete to determine if S can be transformed to S' within the cost m.*

2.2   Approximate EDM

Instead of the exact EDM, an almost linear-time algorithm for approximate EDM was proposed [10]. This approximation algorithm is referred to as edit-sensitive parsing (ESP) algorithm consisting of two phases. One phase is to partition a string into blocks of length two or three in a deterministic manner, and the other phase is to assign consistent names to all blocks. By repeating these phases alternately, a parsing tree of the string is constructed.

These two tasks are alternately executed for each level of T_S as follows. In the parsing phase, ESP partitions $S = s_1 s_2 \cdots s_\ell$ such that $2 \le |s_i| \le 3$ by *alphabet reduction*. Then, the algorithm creates the parent of each s_i in T_S. This means that any internal node of T_S has two or three children. The partition of S can be decided offline, that is, no party needs to communicate for the partition.

---

[2] BGN cannot treat negative integers; therefore, we need another HE system that allows the additive operation.

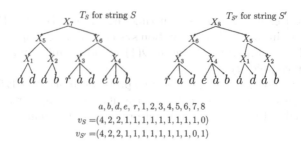

$$a, b, d, e, r, 1, 2, 3, 4, 5, 6, 7, 8$$
$$v_S = (4, 2, 2, 1, 1, 1, 1, 1, 1, 1, 1, 1, 1, 0)$$
$$v_{S'} = (4, 2, 2, 1, 1, 1, 1, 1, 1, 1, 1, 1, 0, 1)$$

Fig. 1. Example of ESP for the strings S and S'. The ESP trees T_S and $T_{S'}$ are constructed by the shared naming function for all internal nodes. After the construction of T_S and $T_{S'}$, the corresponding characteristic vectors v_S and $v_{S'}$ are computed offline. Finally, the unknown $d_*(S, S')$ is approximated by $L_1(v_S, v_{S'}) = 2$.

Thus, we omit the partitioning phase (See [10] for the detail of the alphabet reduction).

Next, in the naming phase, ESP assigns a *consistent* label $L(s_i)$ to the parent of s_i such that $L(s_i) = L(s_j)$ iff $s_i = s_j$. This process is iterated for the resulting $S = L(s_1) \cdots L(s_\ell)$ until $|S| = 1$. In level-wise labeling, if a node is labeled by L at the level i, the name (i, L) is assigned to the node, where $L \in \Sigma$ for $i = 0$. Then, we obtain a consistent T_S: for any two nodes p, q in T_S, $L(p) = L(q)$ iff the sequence of the labels of the children of p and q are equal. By sharing a naming function between S and S', we can construct consistent T_S and $T_{S'}$.

Let $N = \max\{|S|, |S'|\}$ and n be the number of distinct names in the ESP trees T_S or $T_{S'}$. T_S defines a characteristic vector v_S of the length n consisting of the frequency of each name. Finally, we obtain $L_1(v_S, v_{S'})$ as an approximate $d(S, S')$, as shown in Theorem 2.

Theorem 2 ([10]). *Let v_S be the characteristic vector of S such that $v_S[k]$ is the frequency of the name k in T_S. Then, $(1/2)d_*(S, S') \leq L_1(v_S, v_{S'}) = O(d_*(S, S') \lg^* N \lg N)$, where $L_1(v_S, v_{S'}) = \sum_{1 \leq k \leq n} |v_S[k] - v_{S'}[k]|$.*

Figure 1 shows an example of consistent T_S and $T_{S'}$ with the characteristic vectors. The partition of S and S' are executed offline; therefore, the problem of securely computing $d_*(S, S')$ is reduced to two protocols for sharing the naming function and computing $L_1(v_S, v_{S'})$ based on the security of HE.

3   Two-Party Protocol for EDM

3.1   Homomorphic Encryption

In this section, we review two HE systems. The BGN encryption [7] is used for consistent labeling (the first protocol) because a single multiplication and a sufficient number of additive operations over two encrypted integers are required. However, the second protocol for the L_1-distance requires only the additive operation, but BGN cannot treat negative integers. Then, we use the Paillier encryption system [18] admitting additive operations for negative integers.

BGN Encryption System. The BGN encryption system allows addition and multiplication over ciphertexts. For $m_1 \in \{1, \ldots, N\}$ and $m_2 \in \{1, \ldots, N\}$ and their ciphertexts C_1 and C_2, the ciphertexts of $m_1 + m_2$ and $m_1 m_2$ can be computed directly from C_1 and C_2 without knowing m_1 and m_2, if $m_1 + m_2 \leq N$ and $m_1 m_2 \leq N$.

For large primes q_1 and q_2, the encryption system is based on two multiplicative cyclic groups $\mathbb{G}$ and $\mathbb{G}'$ of the order $q_1 q_2$, two generators g_1 and g_2 of $\mathbb{G}$, an inverse function $(\cdot)^{-1} : \mathbb{G} \to \mathbb{G}$, and a bihomomorphism $e : \mathbb{G} \times \mathbb{G} \to \mathbb{G}'$. By definition, $e(\cdot, x)$ and $e(x, \cdot)$ are group homomorphisms for any $x \in \mathbb{G}$. Furthermore, we assume that both $(\cdot)^{-1}$ and e can be computed in polynomial time in terms of the security parameter $\log_2 q_1 q_2$. Such $(\mathbb{G}, \mathbb{G}', g_1, g_2, (\cdot)^{-1}, e)$ can be generated, for example, by letting $\mathbb{G}$ be a subgroup of a supersingular elliptic curve and e a Tate paring [7]. The BGN encryption scheme is described as follows.

Key Generation: Generate two sufficiently large primes q_1 and q_2 at random and $(\mathbb{G}, \mathbb{G}', g_1, g_2, (\cdot)^{-1}, e)$ for $q_1 q_2$ by the means stated above. Choosing two random generators g, u of $\mathbb{G}$, set $h = u^{q_2}$, and let N be a positive integer bounded above by a polynomial function of the security parameter $\log_2 p_1 p_2$. Then, the public key is $PK = (p_1 p_2, \mathbb{G}, \mathbb{G}', e, g, h, N)$ and the private key is $SK = q_1$.

Encryption: Encrypt a message $m \in \{0, \ldots, N\}$ using PK and a random $r \in \mathbb{Z}_n$ by $C = g^m h^r \in \mathbb{G}$. C is the resulting ciphertext.

Decryption: Find m such that $C^{q_1} = (g^m h^r)^{q_1} = (g^{q_1})^m$ using a polynomial time algorithm. An algorithm whose time complexity is $O(\sqrt{N})$ is known in the literature.

Homomorphic Properties: For two ciphertexts $C_1 = g^{m_1} h^{r_1}$ and $C_2 = g^{m_2} h^{r_2}$ in $\mathbb{G}$ of the messages m_1 and m_2, anyone can encrypt $m_1 + m_2$ and $m_1 m_2$ directly from C_1 and C_2 without knowing m_1 and m_2.

Addition: $C = C_1 C_2 h^r = (g^{m_1} h^{r_1})(g^{m_2} h^{r_2}) h^r = g^{m_1 + m_2} h^{r_1 + r_2 + r}$ gives an encryption of $m_1 + m_2$.

Multiplication: $C = e(C_1, C_2) h^r \in \mathbb{G}'$ gives an encryption of $m_1 m_2$, because

$$
\begin{aligned}
C^{q_1} &= e(C_1, C_2) \\
&= [e(g_1, g_2)^{m_1 m_2} e(g_1, g_2)^{q_2 m_1 r_1} e(g_2, g_1)^{q_2 m_2 r_1} e(g_1, g_2)^{q_2 r}]^{q_1} \\
&= (e(g_1, g_2)^{q_1})^{m_1 m_2}.
\end{aligned}
$$

To decrypt C, we compute $m_1 m_2$ from $(g(g_1, g_2)^{q_1})^{m_1 m_2}$ and $e(g_1, g_2)^{q_1}$.

Paillier Encryption System. The Paillier encryption system is described as follows.

Key Generation: Generate two primes p and q at random and let $N = pq$. Then, the public key is $PK = N$, and the private key is $SK = (p, q)$.

Encryption: Encrypt a message $m \in \mathbb{Z}_N$ using PK and a random $r \in \mathbb{Z}_{N^2}$ by $C = (1 + N)^m r^N \mod N^2$. C is the resulting ciphertext.

Decryption: For $\lambda = \mathrm{lcm}((p-1),(q-1))$, $C^\lambda \equiv (1+N)^{\lambda m} r^{N\lambda} \equiv 1 + \lambda m N \bmod N^2$ holds. Hence, we can obtain m by $m = [(C^\lambda - 1)/N] \cdot \lambda^{-1} \bmod N$.

Homomorphic Properties: For two ciphertexts $C_1 = (1 + N)^{m_1} r_1^N \bmod N^2$ and $C_2 = (1 + N)^{m_2} r_2^N \bmod N^2$ of messages m_1 and m_2, anyone can encrypt $m_1 + m_2$ directly from C_1 and C_2 without knowing m_1 and m_2. In fact, $C_1 C_2 \equiv (1 + N)^{m_1 + m_2} (r_1 r_2)^N \bmod N^2$ is a ciphertext of $m_1 + m_2$.

3.2 Protocol for Consistent Labeling

Two parties consisting of $\mathcal{A}$ and $\mathcal{B}$ possess their respective strings $S_\mathcal{A}$ and $S_\mathcal{B}$. Let $[S_\mathcal{A}]$ and $[S_\mathcal{B}]$ be the set of blocks of length two or three, respectively, which are partitioned by an iteration of ESP. At each level of ESP, they want to assign a consistent label L to any of the blocks s and s' such that $L(s) = L(s')$ iff $s = s'$. Let $\pi([S])$ be a permutation of $[S]$ for a string S. For example, if $[S] = \{aa, ab, aab, aba, bbb\}$, we can choose (aab, aa, bbb, ab, aba) as $\pi([S])$. Taking one $\pi = \pi([S_\mathcal{A}] \cup [S_\mathcal{B}])$, we can define a consistent L by the relationship $k = L(s)$ if $\pi[k] = s$ for any block s. However, in a secure protocol, an explicit representation of $[S_\mathcal{A}] \cup [S_\mathcal{B}]$ is not permitted. To securely compute such a consistent L, a sharable L must satisfy the following claim.

Claim. For each $x \in [S_\mathcal{A}]$, $\mathcal{A}$ can obtain a consistent $L(x)$ on $[S_\mathcal{A}] \cup [S_\mathcal{B}]$ and cannot know whether $x \in [S_\mathcal{B}]$ or not. For each $x \in [S_\mathcal{B}]$, $\mathcal{B}$ can obtain a consistent $L(x)$ on $[S_\mathcal{A}] \cup [S_\mathcal{B}]$ and cannot know whether $x \in [S_\mathcal{A}]$ or not.

We give a secure two-party protocol for the consistent labeling in Algorithms 1 and 2, where Algorithm 1 is a subroutine of Algorithm 2. In this protocol, we employ the BGN encryption system admitting unlimited number of additions and depth-one multiplications on encrypted integers. We assume a special symbol $ not appearing in any string to generate dummy strings.

Here, we outline Algorithm 1 executed between the parties $\mathcal{A}$ and $\mathcal{B}$ possessing the sets $[S_\mathcal{A}]$ and $[S_\mathcal{B}]$ of their respective blocks. The task is to jointly assign a consistent name for any block satisfying the required claim. First, the parties add a sufficient number of random dummy blocks into $[S_\mathcal{A}]$ and $[S_\mathcal{B}]$ and compute $\pi_\mathcal{A}$ and $\pi_\mathcal{B}$, where $\pi_\mathcal{A}$ is a permutation of $[S_\mathcal{A}]$ and $\pi_\mathcal{B}$ is analogous. Second, they jointly compute an encrypted random permutation $E_\mathcal{A}(\pi)$ of the multiset $[S_\mathcal{A}] \cup [S_\mathcal{B}]$ by using $\mathcal{A}$'s public key and only $\mathcal{B}$ possesses $E_\mathcal{A}(\pi)$. At this time, they do not know how the multiset $[S_\mathcal{A}] \cup [S_\mathcal{B}]$ was shuffled, that is, they do not know which elements of π originally belonged to their respective sets while decrypting $E_\mathcal{A}(\pi)$. After this preprocessing, they jointly compute a consistent label for each private block. The name of a block w is given as the position k such that $\pi[k] = w$ for the two cases when w appear in π exactly once or twice. (There are only two cases because $\mathcal{A}$ and $\mathcal{B}$ are *semi-honest*; therefore, they enquire about their private blocks only.) Using the property of the BGN encryption system, for each case, $\mathcal{B}$ can compute the encrypted leftmost position $E_\mathcal{A}(k)$ that satisfies $\pi[k] = w$. Receiving $E_\mathcal{A}(k)$, $\mathcal{A}$ obtains the consistent name k without revealing whether $w \in [S_\mathcal{B}]$ or not. For labeling $\mathcal{B}$'s private

---

Algorithm 1 for securely labeling of $[S_\mathcal{A}] \cup [S_\mathcal{B}]$ (subroutine of Algorithm 2)

Input: Two sets $[S_\mathcal{A}]$ and $[S_\mathcal{B}]$ of the blocks possessed by the $\mathcal{A}$ and $\mathcal{B}$ parties, respectively, and the inputs. Let \$ be the special symbol that is not in any string.
Output: A consistent label $L(Q)$ for any query $Q \in [S_\mathcal{A}] \cup [S_\mathcal{B}]$ is the output.
Encryption: $\mathcal{A}$ and $\mathcal{B}$ use their respective public key encryptions $E_\mathcal{A}()$ and $E_\mathcal{B}()$ based on BGN.
Notation: For $X = (x_1, \ldots, x_n)$, $Y = (y_1, \ldots, y_n)$, and encryption scheme $E()$, let $E(X) = (E(x_1), \ldots, E(x_n))$, $X \oplus Y = (x_1 + y_1, \ldots, x_n + y_n)$, and $X \ominus Y = (x_1 - y_1, \ldots, x_n - y_n)$.

Preprocessing
Step1: $\mathcal{A}$ and $\mathcal{B}$ randomly choose $a, b \in \mathbb{N}$ respectively, and randomly generate the dummy blocks r_i ($1 \le i \le a$) and r'_j ($1 \le j \le b$) containing at least one \$. The party chooses random permutations $\pi_\mathcal{A}$ of $[S_\mathcal{A}] \cup \{r_i \mid 1 \le i \le a\}$ and $\pi_\mathcal{B}$ of $[S_\mathcal{B}] \cup \{r'_j \mid 1 \le j \le b\}$.
Step2: $\mathcal{B}$ sends $E_\mathcal{B}(\pi_\mathcal{B})$ to $\mathcal{A}$ by using $\mathcal{B}$'s public key.
Step3: $\mathcal{A}$ generates $E_\mathcal{B}(\pi_1)$ randomly by merging $E_\mathcal{B}(\pi_\mathcal{B})$ and $E_\mathcal{B}(\pi_\mathcal{A})$, computes $E_\mathcal{B}(\pi_1 \oplus R)$ with random $R[i]$ ($1 \le i \le n = |\pi_\mathcal{A}| + |\pi_\mathcal{B}|$), and sends $E_\mathcal{B}(\pi_1 \oplus R)$ and $E_\mathcal{A}(R)$ to $\mathcal{B}$ by using $\mathcal{A}$'s public key.
Step4: $\mathcal{B}$ computes $E_\mathcal{A}(\pi_1) = E_\mathcal{A}(\pi_1 \oplus R \ominus R)$ and generates a random permutation $E_\mathcal{A}(\pi_2) = [E_\mathcal{A}(\alpha_1), \ldots, E_\mathcal{A}(\alpha_n)]$ of $E_\mathcal{A}(\pi_1)$. For simplicity, let $E_\mathcal{A}(\pi)$ denote $E_\mathcal{A}(\pi_2)$.

Labeling for $Q \in [S_\mathcal{A}]$
Step1: $\mathcal{A}$ sends $E_\mathcal{A}(Q)$ to $\mathcal{B}$.
Step2: $\mathcal{B}$ computes $E_\mathcal{A}(\pi \ominus Q) = [E_\mathcal{A}(\alpha_1 - Q), \ldots, E_\mathcal{A}(\alpha_n - Q)]$.
Step3: $\mathcal{B}$ obtains $E_\mathcal{A}(\pi') = [E_\mathcal{A}(a_1), \ldots, E_\mathcal{A}(a_n)]$ by **Sub-protocol**$[E_\mathcal{A}(\pi \ominus Q)]$.
Step4: $\mathcal{B}$ computes

$$E_\mathcal{A}(b_i) = \begin{cases} E_\mathcal{A}(a_1) & \text{if } i = 1 \\ E_\mathcal{A}(a_i(1 - (a_1 + \cdots + a_{i-1}))) & \text{if } 2 \le i \le n. \end{cases}$$

Step5: $\mathcal{B}$ sends $E_\mathcal{A}(q = b_1 + 2b_2 + \cdots + nb_n)$.
Step6: $\mathcal{A}$ decrypts $q = L(Q)$.

Labeling for $Q \in [S_\mathcal{B}]$
Step1: $\mathcal{B}$ obtains $E_\mathcal{A}(q)$ similarly as above.
Step2: $\mathcal{B}$ sends $E_\mathcal{A}(q + r)$ to $\mathcal{A}$ choosing a random r.
Step3: $\mathcal{A}$ decrypts $q + r$ and sends $E_\mathcal{B}(q + r)$ to $\mathcal{B}$.
Step4: $\mathcal{B}$ decrypts $q = L(Q)$.

Sub-protocol$[E_\mathcal{A}(\pi \ominus Q) = E_\mathcal{A}(\alpha_1 - Q), \ldots, E_\mathcal{A}(\alpha_n - Q)]$
$\mathcal{B}$ randomly chooses $a, b \in \mathbb{N}$, $\alpha'_1, \ldots, \alpha'_b \in \{1, \ldots, N\}$ and let $m = n + a + b$.
$\mathcal{B}$ sends a random permutation $E_\mathcal{A}(\beta) = [E_\mathcal{A}(\beta_1), \ldots, E_\mathcal{A}(\beta_m)]$ of the sequence

$$E_\mathcal{A}\left((\alpha_1 - Q), \ldots, (\alpha_n - Q), \overbrace{0, \ldots, 0}^{a}, \alpha'_1, \ldots, \alpha'_b \right) \text{ to } \mathcal{A}.$$

$\mathcal{A}$ decrypts $E_\mathcal{A}(\beta)$ and sends $E_\mathcal{A}(\gamma)$ to $\mathcal{B}$ where $\gamma_i = 1$ if $\beta_i = 0$ and $\gamma_i = 0$ otherwise. $\mathcal{B}$ obtains $E_\mathcal{A}(\pi')$ choosing all $E_\mathcal{A}(\gamma_i)$ corresponding to $E_\mathcal{A}(\alpha_i)$.

Algorithm 2 for securely constructing consistent T_A and T_B

Input: Two strings $S_A, S_B \in \Sigma^N$ possessed by the parties A and B, respectively, and the inputs.

Output: The consistent names for all nodes in T_A and T_B are the outputs.

Step1: A partitions $S_A = s_1 \cdots s_x$ and B partitions $S_B = s_1' \cdots s_y'$ by ESP offline, and they make $[S_A]$ and $[S_B]$, respectively.

Step2: A and B jointly compute $L(Q)$ by Algorithm 1 for each $Q \in [S_A] \cup [S_B]$.

Step3: A transforms $s_1 \cdots s_x$ into $L(s_1) \cdots L(s_x)$, and B similarly transforms $s_1' \cdots s_y'$ into $L(s_1') \cdots L(s_y')$. A and B repeat the above process until $|S_A| = |S_B| = 1$ for the renewal $S_A = L(s_1) \cdots L(s_x)$ and $S_B = L(s_1') \cdots L(s_y')$.

Step4: A defines the name $(\ell, L(s_i))$ for any $s_i \in [S_A]$ in ℓ-th iteration of ESP, and B similarly defines the names for $[S_B]$.

---

blocks, instead of $E_A(k)$, B sends $E_A(k+r)$ with a random integer r to A. After decrypting $k + r$, A sends $E_B(k + r)$ to B using B's public key. Thus, A and B can securely share the consistent name k for any $w \in [S_A] \cup [S_B]$.

By Algorithm 1 outlined above, we can securely assign a consistent name for each node at the same level of the ESP trees T_A and T_B for their private strings. By iterating this based pm the process level, we can obtain consistent derivation trees. The pseudo code is described in Algorithm 2. Next, we show the correctness and complexity of the protocol.

Theorem 3. *Algorithm 1 satisfies the required claim for consistent labeling, and is secure against semi-honest A and B.*

Proof. Note that π in the preprocessing contains any $Q \in [S_A] \cap [S_B]$ exactly twice and any other $Q \in [S_A] \cup [S_B]$ exactly once. Neither r_i nor r_j' is identical to Q because of the special symbol \$. For a query $Q \in [S_A] \cup [S_B]$, π is transformed to $\pi' = [a_1, \ldots, a_n]$ such that $a_i = 1$ if $\pi[i] = Q$ and $a_i = 0$ otherwise. If $Q \in [S_A] \cap [S_B]$, $a_k = a_{k'} = 1$ for some $k < k'$. Using the depth-one multiplication $a_i(1 - (a_1 + \cdots + a_{i-1}))$ in BGN encryption, $[a_1, \ldots, a_n]$ is transformed to $[b_1, \ldots, b_n]$ such that $b_k = 1$ and $b_i = 0$ for all $i \neq k$. If $Q \notin [S_A] \cap [S_B]$, exactly one $a_k = 1$ is stable for the operation. Thus, the labeling determined by $L(Q) = k$ iff $b_k = 1$ satisfies the required claim.

π is a random sequence consisting of expected $O(n)$ zeros and expected $O(n)$ non-zeros at each iteration of ESP; therefore, semi-honest A and B cannot learn anything from the information $L(Q) = k$ under the semantic security of BGN encryption. □

Theorem 4. *Assuming $a + b = O(n)$ for the security parameters a and b in Algorithm 1, the round and communication complexity of Algorithm 2 is $O(\lg N)$ and $O(n \lg n \lg N)$ bits, respectively, where n is the number of labels generated.*

Proof. The round complexity of Algorithm 1 is $O(1)$; therefore, the total round complexity is bounded by the number of function calls from Algorithm 2. Note that for the partition $S_A = s_1 \cdots s_x$ in each iteration of ESP, the resulting string

$L(s_1) \cdots L(s_x)$ satisfies $|L(s_1) \cdots L(s_x)| \leq |S_{\mathcal{A}}|/2$, that is, the depth of ESP for the original strings is $O(\lg N)$ for the length N of the string.

Next, we analyze the communication complexity. The size of the transmitted sequence at each execution of Algorithm 1 is $O(n \lg n)$ bits. The depth of ESP is bounded by $O(\lg N)$; therefore, the total communication in Algorithm 2 is at most $O(n \lg n \lg N)$ bits.   □

3.3   Protocol for L_1-Distance Computation

Algorithm 3 for securely computing $L_1(v_{\mathcal{A}}, v_{\mathcal{B}})$

Input: The characteristic vectors $v_{\mathcal{A}} = (p_1, \ldots, p_n)$ and $v_{\mathcal{B}} = (q_1, \ldots, q_n)$ possessed by $\mathcal{A}$ and $\mathcal{B}$, respectively, are inputs.
Output: $L_1(v_{\mathcal{A}}, v_{\mathcal{B}})$ to be received by $\mathcal{C}$ is the output.
Encryption: $\mathcal{A}$ and $\mathcal{B}$ use $\mathcal{A}$'s public key $E_{\mathcal{A}}()$ based on Paillier encryption.

Step1: $\mathcal{A}$ sends $E_{\mathcal{A}}(v_{\mathcal{A}}) = [E_{\mathcal{A}}(p_1), \ldots, E_{\mathcal{A}}(p_n)]$ to $\mathcal{B}$ using $\mathcal{A}$'s public key.
Step2: $\mathcal{B}$ randomly generates $r_1, \ldots, r_n \in \left(-\frac{N}{2\delta}, \frac{N}{2\delta}\right)$, $a, b \in \mathbb{N}$ and $s_1, \ldots, s_b \in (0, N)$. Let $m = n + a + b$.
Step3: $\mathcal{B}$ randomly chooses a permutation $(t_1, \ldots, t_m)$ of the sequence

$$E_{\mathcal{A}}\left((r_1(p_1 - q_1) \bmod N), \ldots, (r_n(p_n - q_n) \bmod N), \overbrace{0, \ldots, 0}^{a}, s_1, \ldots, s_b\right)$$

and sends $(t_1, \ldots, t_m)$ to $\mathcal{A}$.
Step4: $\mathcal{A}$ decrypts $(t_1, \ldots, t_m)$ and obtains $(u_1, \ldots, u_m)$.
Step5: $\mathcal{A}$ sends $(E_{\mathcal{A}}(|u_1|_N), \ldots, E_{\mathcal{A}}(|u_m|_N))$ to $\mathcal{B}$.
Step6: $\mathcal{B}$ chooses the subsequence corresponding to $E_{\mathcal{A}}(|r_1(p_1 - q_1)|_N), \ldots, E_{\mathcal{A}}(|r_n(p_n - q_n)|_N)$ from $E_{\mathcal{A}}(|u_1|_N), \ldots, E_{\mathcal{A}}(|u_m|_N)$.
Step7: $\mathcal{B}$ sends $E_{\mathcal{A}}(\sum_{i=1}^{n} |r_i(p_i - q_i)|_N |r_i|^{-1} \bmod N)$ to $\mathcal{A}$, where $|r_i|^{-1}$ is the inverse of $|r_i| \pmod{N}$.
Step8: $\mathcal{A}$ obtains $L_1(v_{\mathcal{A}}, v_{\mathcal{B}}) = \sum_{i=1}^{n} |r_i(p_i - q_i)|_N |r_i|^{-1} \bmod N$.

For the ESP trees T_A and T_B, the parties $\mathcal{A}$ and $\mathcal{B}$ compute the characteristic vectors $v_{\mathcal{A}} = (p_1, \ldots, p_n)$ and $v_{\mathcal{B}} = (q_1, \ldots, q_n)$ offline, where p_i and q_i are the frequencies of the names i in T_A and T_B, respectively.

For $x \in \mathbb{N}$, we define the following function $|x|_N$ for computing the absolute value of $x \pmod{N}$.

$$|x|_N = \begin{cases} x \bmod N & \text{if } x \bmod N \in [0, N/2) \\ N - (x \bmod N) & \text{if } x \bmod N \in (N/2, N) \end{cases}$$

We assume that there exist $N, \delta \in \mathbb{N}$ such that for any $p_i, q_i \in [0, \delta]$ and $1 \leq i \leq n$, $p_i - q_i \in (-\delta, \delta)$ and $n\delta < N$ hold.

We outline Algorithm 3 for securely computing $L_1(v_A, v_B)$. In this protocol, we assume the Pailler encryption system because this algorithm needs only the additive operation. First, the party A sends its encrypted vector $E_A(v_A)$ to the party B, and B computes $E_A(r_i(p_i - q_i))$ for each ith element p_i of v_A (resp., q_i of v_B) by choosing the random integers r_i. B generates a random permutation $(t_1, \ldots, t_m)$ consisting of all $E(r_i(p_i - q_i))$ and a sufficient number of encrypted dummys. Next, A decrypts $(t_1, \ldots, t_m)$ to a sequence of palintexts, $(u_1, \ldots, u_m)$ by using his private key and transforms each member u_i to $|u_i|_N$ based on the above definition. Then, A sends $E_A(|u_1|_N, \ldots, |u_m|_N)$ to B. B knows all positions corresponding to the dummy; therefore, B can choose all correct elements in $E_A(|u_1|_N, \ldots, |u_m|_N)$. B also knows the inverse r^{-1} for each r_i; therefore, B can negate the factor r_i as shown in Theorem 5. Finally, B sends the required L_1-distance $E_A(\sum_{i=1}^{n} |r_i(p_i - q_i)|_N |r_i|^{-1} \bmod N)$ to A without revealing their private information.

Theorem 5. *The protocol in Algorithm 3 exactly computes $L(v_A, v_B)$ and is secure against semi-honest A and B.*

Proof. We first show the correctness of the protocol. A receives $\sum_{i=1}^{n} |r_i(p_i - q_i)|_N |r_i|^{-1} \bmod N$. We show this is identical to $L(v_A, v_B) = \sum_{i=1}^{n} |p_i - q_i|$. Note that $r_i(p_i - q_i) \in \left(-\frac{N}{2}, \frac{N}{2}\right)$ for $r_i \in \left(-\frac{N}{2\delta}, \frac{N}{2\delta}\right)$. In case $r_i(p_i - q_i) \in \left(0, \frac{N}{2}\right)$, we obtain $|r_i(p_i - q_i)|_N = r_i(p_i - q_i) = |r_i||p_i - q_i|$. In case $r_i(p_i - q_i) \in \left(-\frac{N}{2}, 0\right)$, we obtain $|r_i(p_i - q_i)|_N = N - (N + r_i(p_i - q_i)) = -r_i(p_i - q_i) = |r_i||p_i - q_i|$. Therefore, since $n\delta < N$, $\sum_{i=1}^{n} |r_i(p_i - q_i)|_N |r_i|^{-1} \bmod N = \sum_{i=1}^{n} |p_i - q_i| \bmod N = \sum_{i=1}^{n} |p_i - q_i|$.

Next, we prove the security of the algorithm. A cannot learn anything about q_i because $(u_1, \ldots, u_m)$ is a random sequence consisting of expected $O(n)$ zeros and expected $O(n)$ non-zeros in $(0, N)$ including $r_i(p_i - q_i) \bmod N$. However, B obtains only $E_A(p_i)$ and $E_A(|r_i(p_i - q_i)|_N)$ generated by $E_A(r_i(p_i - q_i) \bmod N)$. Assuming that the HE algorithm E has indistinguishability under chosen-plaintext attack (IND-CCA secure), $E_A(p_i)$ and $E_A(|r_i(p_i - q_i)|_N)$ are independently distributed in $(0, N)$, that is, B cannot learn anything about p_i. Thus, semi-honest A and B cannot learn anything except $L_1(v_A, v_B)$. $\square$

Theorem 6. *Assuming $a + b = O(n)$ for the security parameters a and b in Algorithm 3, the round and communication complexity of Algorithm 3 for computing $L_1(v_A, v_B)$ are $O(1)$ and $O(n \lg N)$ bits, respectively, where $|v_A| = |v_B| = n$, $n\delta < N$, and $v_A[i], v_B[i] \in [0, \delta)$.*

Proof. Clearly, there are at most 2 rounds between A and B. The largest transmitted message in the rounds is $(t_1, \ldots, t_m)$ in which each t_i is of at most $\lg N$ bits. Thus, we obtain the complexity. $\square$

4   Comparison and Empirical Results

We compare the performance of our protocols with other related methods from the perspective of the number of required encryptions, decryptions, and homomorphic operations over ciphertexts.

Table 1. Complexity analysis for naming protocols: The number of encryptions (#Enc), decryptions (#Dec), and additive and multiplicative operations (#A&M). Here, a multiplication means a single operation for getting $E(cm)$ from $E(m)$ and the plaintext c.

Quantity	Ours	BD [22]	GT-SCOT [6]
#Enc	$O(n \lg N)$	$O(n \lg^2 N)$	$O(n \lg^2 N)$
#Dec	$O(n \lg N)$	$O(n \lg N \lg \lg N)$	$O(n \lg N)$
#A&M	$O(n \lg N)$	$O(n \lg^2 N)$	$O(n \lg^2 N)$

Table 2. Complexity analysis for L_1-distance protocols: the number of encryptions (#Enc), decryptions (#Dec), and additive and multiplicative operations (#A&M) for the L_1 computing. Possibly, $n = \Omega(N)$ in a worst-case scenario for the length n of the vectors.

Quantity	Ours	Exact (UE) [21]	Approximate [21]
#Enc	$O(n)$	$O(nN)$	$O(n \lg N)$
#Dec	$O(n)$	$O(1)$	$O(1)$
#A&M	$O(n)$	$O(nN)$	$O(n \lg N)$

Table 1 shows the performance for the first problem: the consistent labeling for $T_{\mathcal{A}}$ and $T_{\mathcal{B}}$. The bit-decomposition [22] and GT-SCOT [6] require bit-wise encryption for each integer; therefore, a naive protocol based on these methods incurs a larger overhead. However, our method directly encrypts each integer and uses it as it is.

Table 2 shows the performance for the second problem: the L_1-distance of $v_{\mathcal{A}}$ and $v_{\mathcal{B}}$. An exact and approximate computation is proposed in [21]. Since $n = \Omega(N)$ in a worst-case scenario (e.g., consider a string consisting of no frequent symbol), both methods are not practical for large strings. Our algorithm significantly reduces the overhead by using the proposed protocol for computing the absolute value. Moreover, we evaluate the performance of the proposed algorithms for securely computing EDM. We implemented Algorithms 1 and 2. For comparison, we also implemented native methods based on bit-decomposition (BD) [22] for consistent labeling and unary encoding ([21], Table 2) for computing the L_1 distance.

These algorithms were implemented in C++ based on a library available from https://github.com/shaih/HElib. We compiled the programs with Apple LLVM version 8.0.0 (clang-800.0.42.1) on Mac OS X 10.11.6 and examined the performance on 1.6 GHz Intel Core i5 and 4 GB 1600 MHz DDR3 memory.

Table 3 shows the empirical results of these algorithms for $n \in \{10, 100, 1000\}$ where n is the number of different labels generated in the first protocol or the length of the characteristic vector in the second protocol; these labels are identical each other. The size of the integer (plaintext) to be encrypted is at most 32 bits and the key length in each encryption system is set to 256 bits. In Table 3, the total time of the proposed method is the sum of Algorithms 1 and 2. However, the total time of the native method is the sum of BD and Exact. By this experiment, we confirmed that the computation time is significantly improved.

Table 3. Empirical results of secure EDM computation: Algorithms 1 and 2 show the running time in seconds for the proposed consistent labeling and L_1-distance computation, respectively. The total time of the proposed secure EDM computation is the sum of Algorithms 1 and 2. We compare the running time (s) of the naive implementations for the consistent labeling based on the BD and the L_1-distance computation based on the unary encoding (UE) are also reported.

#label (n)	Algorithm 1	BD [22]	Algorithm 2	UE [21]
10	0.200	15.002	4.514	1.090
100	0.223	138.393	4.543	10.355
1000	0.2192	1341.245	8.310	1873.353

5   Conclusion

We proposed a secure two-party protocol for EDM based on HE by solving two problems. The first problem was reducible to securely computing the total order of the multiset $[S_A] \cup [S_B]$ for the sets of phrases shared by the parties A and B. We propose a solution for this problem by assuming a single multiplication of ciphertexts. Whether or not we can construct a protocol with only the additive homomorphic property is an open question. The second problem was reducible to securely computing $|p - q|$. Previous studies used bitwise encryption for p and q; therefore, they required large overheads. Our method can determine the sign of $p - q$ using direct encryption of p and q. However, the technique related to the oblivious RAM [19,25] will potentially improve the preprocessing of our algorithm; this will be an important challenge in the future.

References

1. Akgün, M., Bayrak, A.O., Ozer, B., Sağiroğlu, M.S.: Privacy preserving processing of genomic data: a survey. J. Biomed. Inform. **56**, 103–111 (2015)
2. Atallah, M.J., Kerschbaum, F., Du, W.: Secure and private sequence comparisons. In: WPES, pp. 39–44 (2003)
3. Aziz, M.M.A., Alhadidi, D., Mohammed, N.: Secure and efficient multiparty computation on genomic data. In: IDEAS, pp. 278–283 (2016)
4. Beck, M., Kerschbaum, F.: Approximate two-party privacy-preserving string matching with linear complexity. In: BigData Congress, pp. 31–37 (2013)
5. Belazzougui, D., Zhang, Q.: Edit distance: sketching, streaming, and document exchange. In: FOCS, pp. 51–60 (2016)

6. Blake, I.F., Kolesnikov, V.: Strong conditional oblivious transfer and computing on intervals. In: Lee, P.J. (ed.) ASIACRYPT 2004. LNCS, vol. 3329, pp. 515–529. Springer, Heidelberg (2004). https://doi.org/10.1007/978-3-540-30539-2_36

7. Boneh, D., Goh, E.-J., Nissim, K.: Evaluating 2-DNF formulas on ciphertexts. In: Kilian, J. (ed.) TCC 2005. LNCS, vol. 3378, pp. 325–341. Springer, Heidelberg (2005). https://doi.org/10.1007/978-3-540-30576-7_18

8. Catalano, D., Di Raimondo, M., Faro, S.: Verifiable pattern matching on outsourced texts. In: Zikas, V., De Prisco, R. (eds.) SCN 2016. LNCS, vol. 9841, pp. 333–350. Springer, Cham (2016). https://doi.org/10.1007/978-3-319-44618-9_18

9. Cheon, J.H., Kim, M., Lauter, K.E.: Homomorphic computation of edit distance. In: FCW, pp. 194–212 (2015)

10. Cormode, G., Muthukrishnan, S.: The string edit distance matching problem with moves. ACM Trans. Algorithms 3(1) (2007). Article 2

11. Gentry, C.: Fully homomorphic encryption using ideal lattices. In: STOC, pp. 169–178 (2009)

12. Goldwasser, S., Kalai, Y., Popa, R.A., Vaikuntanathan, V., Zeldovich, N.: Reusable garbled circuits and succinct functional encryption. In: STOC, pp. 555–564 (2013)

13. Hach, F., Numanagić, I., Alkan, C., Sahinalp, S.C.: Scalce: boosting sequence compression algorithms using locally consistent encoding. Bioinformatics 28(23), 3051–3057 (2012)

14. Inan, A., Kaya, S., Saygin, Y., Savas, E., Hintoglu, A., Levi, A.: Privacy preserving clustering on horizontally partitioned data. Data Knowl. Eng. 63(3), 646–666 (2007)

15. Jowhari, H.: Efficient communication protocols for deciding edit distance. In: Epstein, L., Ferragina, P. (eds.) ESA 2012. LNCS, vol. 7501, pp. 648–658. Springer, Heidelberg (2012). https://doi.org/10.1007/978-3-642-33090-2_56

16. Li, M., Chen, X., Li, X., Ma, B., Vitanyi, P.M.B.: The similarity metric. IEEE Trans. Inform. Theory 50(12), 3250–3264 (2004)

17. Maruyama, S., Tabei, Y.: Fully-online grammar compression in constant space. In: DCC, pp. 218–229 (2014)

18. Paillier, P.: Public-key cryptosystems based on composite degree residuosity classes. In: Stern, J. (ed.) EUROCRYPT 1999. LNCS, vol. 1592, pp. 223–238. Springer, Heidelberg (1999). https://doi.org/10.1007/3-540-48910-X_16

19. Patel, S., Persiano, G., Yeo, K.: Recursive orams with practical constructions. Cryptology ePrint Archive, Report 2017/964 (2017)

20. Rane, S., Sun, W.: Privacy preserving string comparisons based on Levenshtein distance. In: WIFS, pp. 1–6 (2010)

21. Rane, S., Sun, W., Vetro, A.: Privacy-preserving approximation of L1 distance for multimedia applications. In: ICME, pp. 492–497 (2010)

22. Samanthula, B.K.K., Chun, H., Jiang, W.: An efficient and probabilistic secure bit-decomposition. In: ACM SIGSAC Symposium on Information, Computer and Communications Security, pp. 541–546 (2013)

23. Shapira, D., Storer, J.A.: Edit distance with move operations. J. Discrete Algorithms 5(2), 380–392 (2007)

24. Starikovskaya, T.: Communication and streaming complexity of approximate pattern matching. In: CPM, pp. 13:1–13:11 (2017)

25. Stefanov, E., et al.: Path oram: an extremely simple oblivious ram protocol. In: CCS, pp. 299–310 (2013)

26. Takabatake, Y., I, T., Sakamoto, H.: A space-optimal grammar compression. In: ESA, pp. 67:1–67:15 (2017)

27. Toft, T.: Constant-rounds, almost-linear bit-decomposition of secret shared values. In: Fischlin, M. (ed.) CT-RSA 2009. LNCS, vol. 5473, pp. 357–371. Springer, Heidelberg (2009). https://doi.org/10.1007/978-3-642-00862-7_24
28. Yao, A.C.: How to generate and exchange secrets. In: FOCS, pp. 162–167 (1986)
29. Zhu, R., Huang, Y.: Efficient privacy-preserving edit distance and beyond. IACR Cryptology ePrint Archive 2017: 683 (2017)

Shared Session SISAP and SPIRE

On the Analysis of Compressed Chemical Fingerprints

Fabio Grandi$^{(\boxtimes)}$

Department of Computer Science and Engineering (DISI),
Alma Mater Studiorum – Università di Bologna, Viale Risorgimento 2,
40136 Bologna, BO, Italy
`fabio.grandi@unibo.it`

Abstract. Chemical fingerprints are binary strings used to represent the distinctive features of molecules in order to efficiently support similarity search of chemical data. In large repositories, chemical fingerprints are conveniently stored in compressed format, although the lossy compression process may introduce a systematic error on similarity measures. Simple correction formulae have proposed by Swamidass and Baldi in [13] to compensate for such an error and, thus, to improve the similarity-based retrieval. Correction is based on deriving estimates for the weight (i.e., number of bits set to 1) of fingerprints before compression from their compressed values. Although the proposed correction has been substantiated by satisfactory experimental results, the way in which such estimates have been derived and the approximations applied in [13] are not fully convincing and, thus, deserve further investigation. In this direction, the contribution of this work is to provide some deeper insight on the fingerprint generation and compression process, which could constitute a more solid theoretical underpinning for the Swamidass and Baldi correction formulae.

1 Introduction

In chemoinformatics, fixed-length binary *fingerprints* [12,13] are an encoding method used for fast executing similarity queries in large repositories of chemical data concerning molecules. Individual bits in a fingerprint may encode the presence or absence (with a 1 or 0 value, respectively) of some paths of atoms and bonds in the molecule or of some complex structural property. Molecule fingerprints can be fast searched and matched against a query pattern using some similarity measure, like the Jaccard/Tanimoto or Tversky measure [11,14]. For instance, in order to identify suitable candidates in drug design programs [2,10], thousands of query fingerprints are screened for similarity against millions of fingerprints stored in a database. In order to cope with their resulting sparsity, long fingerprints are often converted via k-fold compression into shorter ones (e.g., in the commercial fingerprint system Daylight [3]), which are more handy to store and manage and, in particular, less sparse than their uncompressed counterparts. Special index structures may also be used to speed up the fingerprint database search [1].

© Springer Nature Switzerland AG 2018
S. Marchand-Maillet et al. (Eds.): SISAP 2018, LNCS 11223, pp. 243–256, 2018.
https://doi.org/10.1007/978-3-030-02224-2_19

Measures of similarity between fingerprints are based on their overlap (i.e., number of 1 bits in common). Due to sparsity reduction and lossy compression, compressed signatures present a higher overlap than their uncompressed form and such increase in overlap can be considered a systematic error introduced by the compression method. As a consequence, the quality of similarity retrieval would heavily rely on an arbitrary chosen fingerprint length [4]. Fortunately, Swamidass and Baldi showed how this error can be compensated with simple formulae and experimentally validated their approach in [13]. Their similarity compensation approach is based on the estimation of the weight (i.e., number of bits set to 1) of a fingerprint before compression given the weight of the fingerprint after compression. Although their estimation method has been shown to produce fruitful results, it is based on assumptions and approximations which are not fully convincing from a theoretical viewpoint and deserve a deeper insight to justify their validity. Such a deeper insight is the objective of this work.

The rest of the paper is organized as follows. In Sect. 2, notation is introduced and background on the approach in [13] is recalled. Section 3 is devoted to a new more rigorous analysis of the fingerprint generation and compression process, including: derivation of an exact probabilistic model of the mutual dependence between the compressed and uncompressed weights in Subsect. 3.1, a discussion of its applicability and the introduction of a more realistic model for the dependence of the uncompressed weight on the compressed weight in Subsect. 3.2 and an experimental validation of the predictive power of the newly proposed and previous estimation formulae in Subsect. 3.3. Finally, conclusions can be found in Sect. 4.

2   Preliminaries

In this Section, after introducing some notation, we briefly review the approximate analysis presented in [13] of the fingerprint compression process.

Let us start with the introduction of a bit of notation to be used in the following. First of all, if $\mathbf{B}$ is a bit string of length L, we denote by $\mathbf{B}[i]$ the i-th bit of $\mathbf{B}$ ($i \in \{1..L\}$) and by $\mathbf{B}[i : j]$ the substring of $\mathbf{B}$ composed of the consecutive bits from i to j inclusive ($1 \leq i < j \leq L$).

Mainly adopting the symbols used in [13], we consider $\mathbf{X}_*$ to be an uncompressed fingerprint, with length N_*, and $\mathbf{X}$ its compressed version, with length N. The compressed fingerprint $\mathbf{X}$ is obtained by considering $\mathbf{X}_*$ as composed by the concatenation of k *frames* of consecutive $N = N_*/k$ bits each (i.e., the i-th frame $\mathbf{F}_i$ is defined as the bit substring $\mathbf{X}_*[N(i-1)+1, Ni]$, $i \in \{1..k\}$) and folding it by superimposing the frames: $\mathbf{X} = \bigvee_{i=1}^{k} \mathbf{F}_i$. Hence $\mathbf{X}[j] = \bigvee_{i=1}^{k} \mathbf{F}_i[j]$, that is the j-th bit of $\mathbf{X}$ is set to 1 iff in at least one of the frames of $\mathbf{X}_*$ the j-th bit is 1. N_*, N and k are commonly assumed to be powers of 2: N=512 or N=1024 are common choices, with $N_* = 2^n$ with n in the range 15–20 depending on the selected features (e.g., depth of the atom/bond paths) and also on the molecule database size [13]. With A_* and A, we denote the weight of $\mathbf{X}_*$ and $\mathbf{X}$, respectively.

Instead of decomposing the fingerprint $\mathbf{X}_*$ into k N-bit frames $\mathbf{F}_i$ ($i \in \{1..k\}$), we can also decompose it into N k-bit *slices* $\mathbf{S}_j$ ($j \in \{1..N\}$), where $\mathbf{S}_j[i] = \mathbf{F}_i[j]$ ($i \in \{1..k\}$, $j \in \{1..N\}$). Hence, there is a one-to-one correspondence between $\mathbf{S}_j$ and $\mathbf{X}[j]$, such that $\mathbf{X}[j] = \bigvee_{i=1}^{k} \mathbf{S}_j[i]$ and the j-th bit of $\mathbf{X}$ is set to 1 iff at least one of the k bits in the slice $\mathbf{S}_j$ is set to 1; in such a case, we will call it a *qualifying slice*.

2.1 Background on Approximate Analysis

The problem of chemical fingerprint compression has been studied by Swamidass and Baldi in [13]. In order to derive an estimate for the similarity of uncompressed fingerprints from the similarity of compressed fingerprints, the relationship between A_* and A due to the k-fold compression method has to be understood. In particular, a reliable estimate of A_* given the weight A of the compressed fingerprint has to be determined. Considering A_* and A as random variables, the authors of [13] assumed that such an estimate consists of the conditional expected value $\mathrm{E}[A_*|A]$ and introduced an underlying probabilistic model to evaluate it. Their probabilistic approach is based on the approximations and simplifications described in the following.

The first simplification they introduced is the assumption of a *fixed-density* model for the fingerprint compression process, where bits of $\mathbf{X}_*$ have a constant probability $\alpha = A_*/N_*$ to be found set to 1 when they are iteratively selected in order to assemble a slice. The resulting distribution of A given A_* underlying the compression process is binomial, with the probability $1 - (1 - \alpha)^k$ to have a bit set to 1 in $\mathbf{X}$, leading to the estimate:

$$\mathrm{E}[A|A_*] \approx N\left[1 - \left(1 - \frac{A_*}{N_*}\right)^k\right] \tag{1}$$

Then, treating (1) as a deterministic functional dependency between A_* and A, instead of using the Bayes theorem and computing the correct expectation, they inverted such dependency leading to their first estimate of $\mathrm{E}[A_*|A]$:

$$\mathrm{E}[A_*|A] \approx N_*\left[1 - \left(1 - \frac{A}{N}\right)^{1/k}\right] \tag{2}$$

Furthermore, they introduced an asymptotic approximation for (1) (valid for large N_* values), yielding:

$$\mathrm{E}[A|A_*] \approx N\left(1 - e^{-A_*/N}\right) \tag{3}$$

which can be inverted leading to their second estimate of $\mathrm{E}[A_*|A]$ (valid when A is not close to N):

$$\mathrm{E}[A_*|A] \approx -N\log\left(1 - \frac{A}{N}\right) \tag{4}$$

The values provided by Eq. (4) have then been used to apply the due correction to the similarity measures of compressed fingerprints. The proposed formulae have been satisfactorily validated by the experiments reported in [13]. However, the derivation of (4) and the justification of its validity leaves something to be desired, which is the starting point of the present work.

3   A New Improved Analysis of Compressed Fingerprints

Our new more rigorous analysis of compressed fingerprints is presented in this Section.

3.1   An Exact Probabilistic Model of the Compression Process

This subsection is devoted to a new and exact probabilistic characterization of the fingerprint compression process. In particular, we begin with applying the γ-transform approach—described in [7] and that we first introduced in [5]—to the analysis of the compression process. The γ-transform approach has been proved very useful for a rapid exact probabilistic characterization of counting problems recurring, for instance, in performance analysis of databases [7], signature files [6] and Bloom filters [8]. In this work, using the γ-transform, we will easily derive the exact probability mass function, expected value and variance of the number of bits A set to 1 in a compressed fingerprint conditioned to the number of bits A_* set to 1 in the uncompressed one.

In a counting experiment where possible outcomes can be selected from a set with cardinality m, the γ-transform $\gamma(y)$ of the probability mass function of the number of outcomes can be evaluated as the probability of selecting outcomes from a subset with cardinality $y \leq m$ only [7, Theorem 3]. The random variable we consider here is the number of qualifying slices in an uncompressed fingerprint conditioned to the number of bits set to 1 in $\mathbf{X}_*$, that is A conditioned to A_*. Owing to the physical meaning of the γ-transform recalled above, and since the selection of bits to be set to 1 is without replacement, we have:

$$\gamma(y) \;=\; \frac{\dbinom{yk}{A_*}}{\dbinom{N_*}{A_*}} \tag{5}$$

as $\binom{N_*}{A_*}$ is the total number of ways in which we can select A_* bits in $\mathbf{X}$ and $\binom{yk}{A_*}$ is the number of ways in which we can select (without replacement) A_* bits from y slices of $\mathbf{X}$ only.

Hence, using formulae (6), (13) and (14) of [7], we can derive in a straightforward way from $\gamma(y)$ the probability mass function, expected value and variance of A conditioned to A_*, respectively, as:

$$P(A|A_*) = \binom{N}{A} \sum_{j=0}^{A} (-1)^j \binom{A}{j} \frac{\binom{(A-j)k}{A_*}}{\binom{N_*}{A_*}} \tag{6}$$

$$E[A|A_*] = N \left[1 - \frac{\binom{N_* - k}{A_*}}{\binom{N_*}{A_*}} \right] \tag{7}$$

$$\sigma^2_{A|A_*} = N^2 \left[\frac{\binom{N_* - 2k}{A_*}}{\binom{N_*}{A_*}} - \frac{\binom{N_* - k}{A_*}^2}{\binom{N_*}{A_*}^2} \right]$$

$$+ N \left[\frac{\binom{N_* - k}{A_*}}{\binom{N_*}{A_*}} - \frac{\binom{N_* - 2k}{A_*}}{\binom{N_*}{A_*}} \right] \tag{8}$$

The formulae above agree with the ones we derived in [5,7] for the "group inclusion problem". The probability mass function (6) is the same as derived by Swamidass and Baldi in [13, App. 1], whereas no expression for the exact expected value (7) and variance (8) has been found in their studies. Notice that knowing the variance expression (which would be quite hard to derive from (6) without the help of the γ-transform theory) is useful to test the goodness of approximations (e.g., including derivation of (2) from (1)) treating the expected value as if it was a deterministic value, since it allows to evaluate how the distribution is concentrated around the expected value, for instance by means of the Tchebycheff inequality.

The expected value (7) can also be directly computed by considering that $\binom{N_* - k}{A_*}/\binom{N_*}{A_*}$ is the exact probability that A_* bits are selected in X_* by excluding a given slice, namely the probability that a given slice does not contain any of the A_* bits set to 1. Therefore, (7) represents the expected value of the number of slices containing at least one of the A_* bits set to 1. Notice that determining (7) here is the same problem of estimating in data management the number of disk blocks containing a desired number of records leading to the Yao's formula [15] (where N_*, N, k, A_* play the role, respectively, of the number of records, blocks, records in a block and records sought).

Then, in order to also determine $P(A|A_*)$ from $P(A_*|A)$, the Bayes theorem can be used:

$$P(A|A_*) = \frac{P(A_*|A)P(A)}{P(A_*)} \tag{9}$$

The probability $P(A_*)$ can be easily computed as it is the probability that A_* bits are set to 1 in a random string of N_* bits:

$$P(A_*) = \frac{\binom{N_*}{A_*}}{2^{N_*}} \tag{10}$$

In fact, there are $\binom{N_*}{A_*}$ ways to select (without replacement) the A_* bits to be set to 1, while 2^{N_*} is the total number of ways in which the N_*-bit string can be built.

The probability $P(A)$ is a bit more tricky to evaluate, since it is the probability that A bits are set to 1 in a string of N bits generated via k-fold compression of a random string of N_* bits. By the way, we already have all the elements needed to compute it in an indirect way as follows. From (6) and (10) we can compute the joint probability of A_* and A:

$$P(A_*, A) = P(A|A_*)P(A_*)$$
$$= \frac{\binom{N}{A}}{2^{N_*}} \sum_{j=0}^{A} (-1)^j \binom{A}{j} \binom{(A-j)k}{A_*} \tag{11}$$

from which $P(A)$ can be computed by marginalization:

$$P(A) = \sum_{A_*=0}^{N_*} P(A_*, A)$$
$$= \frac{\binom{N}{A}}{2^{N_*}} \sum_{j=0}^{A} (-1)^j \binom{A}{j} \sum_{A_*=0}^{N_*} \binom{(A-j)k}{A_*} \tag{12}$$

which, by applying two times the binomial theorem, yields:

$$P(A) = \frac{\binom{N}{A} (2^k - 1)^A}{2^{N_*}} \tag{13}$$

However, the $P(A)$ expression (13) can also be explained and directly derived as ratio between the numbers of favorable and total configurations. The number of favorable configurations, among the 2^{N_*} total ones, can be determined as follows. First, in order to produce a compressed fingerprint with A bits set to 1, A of the N slices of $\mathbf{X}_*$ must be qualifying. Second, in a k-bit slice, there are $2^k - 1$ ways of placing 1 bits in order to build a qualifying one (only the configuration with all 0 bits must be excluded). Hence, considering altogether the A qualifying slices, which can be built independently on each other, there are $(2^k - 1)^A$ ways to place the 1 bits in them. Since there are $\binom{N}{A}$ ways to choose (without replacement) the qualifying slices in $\mathbf{X}_*$, the total number of favorable configurations is $\binom{N}{A}(2^k - 1)^A$, still yielding (13).

Hence, by substituting (6), (10) and (13) in (9), we finally obtain:

$$P(A_*|A) = \frac{1}{(2^k-1)^A} \sum_{j=0}^{A}(-1)^j \binom{A}{j} \binom{(A-j)k}{A_*}$$ (14)

Using the identity $\binom{n}{k} = \frac{n}{k}\binom{n-1}{k-1}$ and the binomial theorem, with simple manipulations we can compute from (14) the conditional expected value of A_* given A:

$$E[A_*|A] = \sum_{A_*=0}^{N_*} A_* P(A_*|A)$$

$$= \frac{1}{(2^k-1)^A} \sum_{j=0}^{A}(-1)^j \binom{A}{j} \sum_{A_*=0}^{N_*} A_* \binom{(A-j)k}{A_*}$$

$$= kA\frac{2^{k-1}}{2^k-1}$$ (15)

which has a very simple expression showing that its value is proportional to A. In particular, if k is not too small, $E[A_*|A] \approx kA/2$.

Notice that (15) can also be computed directly as follows. Let us focus on a bit of a qualifying slice of $\mathbf{X}_*$ and let p be the probability that such a bit is set to 1. The number of ways in which this can happen is 2^{k-1} (while this bit is set to 1, the values of the other $k-1$ bits can be freely set in 2^{k-1} different ways), whereas the total number of ways a qualifying slice can be built is 2^k-1 (among all the 2^k possible realizations of a slice, we only must exclude the one with all bits set to 0 in order to build a qualifying one). Hence $p = 2^{k-1}/(2^k-1)$. Since all the k bits in a qualifying slice have the same probability p to be set to 1, the expected value of the number of bits set to 1 in a slice is kp. Since each of the A qualifying slices contributes by kp to the expected number of bits set to 1 in $\mathbf{X}_*$ and the non qualifying slices do not contribute, the expected number of bits globally set to 1 in $\mathbf{X}_*$ is kAp, yielding (15).

At this point, we can underline how the exact value (15) of $E[A_*|A]$ is quite different from the "approximate" values (2) or (4), which have been used in [13], and whose adoption was proven though to give satisfactory predictions of A_* given A in an experimental setting. Hence, there is something else concerning the construction of compressed fingerprints which needs to be explained. This is the aim of the next Subsection.

3.2   A Deeper Insight on the Fingerprint Construction Process

Although (15) is the exact expected value of A_* given A, its knowledge is of little use in the application context of chemical fingerprints indeed. The ratio under this statement, first of all, is the fact that (15) is computed allowing the random variable A_* to range from 0 to N_*, which does not take into account the *sparsity* of the uncompressed fingerprints. Hence, in order to find an inverse relationship between A and A_* taking into account fingerprint sparsity, we must consider the maximum value $A_*^{\max}$ of A_* in the fingerprint database before compression and evaluate the average value of A_* computed over all the fingerprints with weight up to $A_*^{\max}$ which give rise to a compressed fingerprint with weight A.

Moreover, this average value cannot be computed simply using the expectation formula (15) with the summation on A_* limited to the first $A_*^{\max}$ terms. Such a computation would be based on the fact that, when a slice is qualifying, all the combinations of 1 bits in it, in a number from 1 to k, can be considered, regardless of what happens in the other slices. Owing to the *random* selection of bits to be set to 1 in the uncompressed fingerprint, their average density must be as much as possible constant over the different slices. Therefore, when there are still non-qualifying slices (i.e., with a local density sticking to 0), it is almost impossible that 1 bits pile up in the qualifying ones. Hence, in a real situation, only uncompressed fingerprints with the lowest possible weights contribute to the determination of A_* given A, as further explained in the following.

Construction Dynamics Exposed. Let us take a closer look to the construction process of an uncompressed fingerprint, where bits are incrementally set to 1 and their number grows from 1 to A_* and let us call Φ the function that maps A_* into the expected value of A given by (7) (i.e., such that $\Phi(A_*) = \mathrm{E}[A|A_*]$). Initially, the number of qualifying slices $\Phi(A_*)$ grows linearly with A_* as it is very likely that the first few bits set to 1 hit different slices, owing to the fact that bits to be set to 1 are randomly chosen in the uncompressed fingerprint (i.e., each of the 0 bits has the same probability to be set to 1). In fact, there is a negligible probability that two of the first few bits to be set to 1 hit the same slice, much lower indeed than the probability that such bits hit different slices (e.g., for the first two bits such probabilities are $1/N$ and $(N-1)/N = 1 - 1/N$, respectively).

When the number of hit slices grows, the probability that the next bit to be set to 1 hit an already qualifying slice grows and the function $\Phi(A_*)$ becomes less than linear. However, as we already observed, while there are still non qualifying slices (i.e., $\Phi(A_*) < N$), it is very likely that the further bits to be set to 1 will hit some of the still non qualifying slices making them qualify rather than pile up in the already qualifying ones.

Let $\overline{A}_*$ be the weight of uncompressed fingerprints that, during the compression process, first saturate the maximum number N of qualifying slices. The continuous function Φ is monotonically growing and asymptotically limited by N and, in order to provide an integral number of qualifying slices, its values must be rounded to the nearest integer. Hence $\overline{A}_* = \min\{A_* \mid \lfloor \Phi(A_*) + 0.5 \rfloor = N\}$ yielding, thanks to monotonicity, $\overline{A}_* = \Phi^{-1}(N - 0.5)$.

While $A_*^{\max} < \overline{A}_*$, none of the fingerprints in the uncompressed database is able to saturate the number of qualifying slices. During the fingerprint construction process, given the number of currently qualifying slices $A < N$, each of them must contain, on average, the minimum possible number m of 1 bits compatible with A (otherwise, additional 1 bits would have hit some of the

$N - A$ still non-qualifying slices rather than accumulate in one of the A already qualifying ones). Hence, also the whole uncompressed fingerprint contains a minimum number $A_* = mA$ of 1 bits, which must be compatible with the constraint $\Phi(A_*) = A$, that is $A_* = \Phi^{-1}(A)$ thanks to the monotonicity of Φ. Notice that, when $A < N$, the effects of the rounding of Φ to the nearest integer are negligible and, thus, we will not evidence them in formulae.

When $A_*^{\max} \geq \overline{A}_*$, some of the fingerprints in the database after compression saturate the number N of slices and their $A_* - \overline{A}_*$ bits, exceeding the $\overline{A}_*$ ones necessary for saturation, are further distributed among slices. Also in such a case, the fingerprints with weight less that $\overline{A}_*$ still contribute to compressed fingerprints with a weight A less than N according to $A = \Phi(A_*)$ and, thus, the inverse relationship $A_* = \Phi^{-1}(A)$ still holds for each $A < N$ (regardless of the $A_*^{\max}$ value). When $A = N$, the uncompressed fingerprints must have a weight including the $\overline{A}_*$ bits necessary to saturate the number of slices plus a number of exceeding bits, which may range from the $(\overline{A}_* + 1)$-th up to the $A_*^{\max}$-th. Hence the number of such exceeding bits is, on average, $(A_*^{\max} - \overline{A}_*)/2$ and, thus, we have an estimate $\overline{A}_* + (A_*^{\max} - \overline{A}_*)/2$ for the average weight of an uncompressed fingerprint giving rise to a compressed weight of N. The resulting expression for the estimation of A_* is, thus, as follows:

$$
A_* \simeq
\begin{cases}
\Phi^{-1}(A) & \text{if } A < N \\
\\
\Phi^{-1}(N - 0.5) + \left[A_*^{\max} - \Phi^{-1}(N - 0.5)\right]/2 & \text{if } A = N
\end{cases}
\tag{16}
$$

Notice that the adoption of $\Phi^{-1}(A)$ as an estimate for A_* given A, which is needed for computing the right similarity measure in compressed fingerprint databases, has been a sort of fortunate choice in [13]. In fact, such a choice was based on two assumptions:

- the needed estimate for A_* given A is $\mathrm{E}[A_* | A]$
- an approximation of $\mathrm{E}[A_* | A]$ can be computed as $\Phi^{-1}(A)$

which we showed that are both incorrect. However, if we short-circuit the two assumptions by cancelling out the intermediate passage through $\mathrm{E}[A_* | A]$, we obtain the correct solution to our estimation problem.

In practice, Eqs. (15) and (16) provide mutually consistent estimations (i.e., about $N_*/2$) of A_* given A only when $A = N$ and $A_*^{\max} = N_*$.

Accurate Evaluation of Φ^{-1}. Continuing our analysis, the next step is to evaluate the inverse of the function Φ. The Eq. (7) defining $\Phi(A_*)$ contains A_* inside binomial coefficients and, thus, cannot be solved in closed form for A_*. Therefore, an invertible approximation must be found for the expression appearing in the right-hand side. A quite natural way to approximate (7) getting rid of the binomial coefficients is to consider selection of bits with replacement instead of selection without replacement.

Equation (7) is based on the probability $\binom{N_* - k}{A_*} / \binom{N_*}{A_*}$ that a given slice is excluded when the A_* bits to be set to 1 are chosen in $\mathbf{X}$ and, considering selection of the bits with replacement instead, can be simply approximated as $(1 - k/N_*)^{A_*}$. The exact probability is also equal to $\binom{N_*-A_*}{k}/\binom{N_*}{k}$, representing the probability of assembling a slice by only choosing fingerprint positions non containing any of the A_* bits set to 1 and, considering selection of bit positions with replacement instead, can be simply approximated as $(1 - A_*/N_*)^k$. These two ways in which the exact probability and its replacement-based approximation can be computed correspond to the two viewpoints implied in what we called *primal* and *dual experiment*, respectively, in [9]. On one hand, the replacement-based approximation of the primal experiment is physically feasible only when $A_* \leq k$ (otherwise, it would allow a 1 bit to be selected in a slice more times than the total number k of bits in it, which is clearly impossible in the real experiment). On the other hand, the replacement-based approximation of the dual experiment is physically feasible only when $k \leq A_*$ (otherwise, it would allow a bit position set to 1 to be selected to build up a slice more times than the total number A_* of 1 bits in the whole fingerprint, which is clearly impossible in the real experiment). Therefore, as we proposed in [9], the two replacement-based approximations involved in the primal and dual experiments can be combined in order to produce a globally feasible (and, thus, globally better) approximation as follows:

$$\frac{\binom{a-c}{b}}{\binom{a}{b}} = \frac{\binom{a-b}{c}}{\binom{a}{c}} \simeq \begin{cases} \left(1 - \frac{c}{a}\right)^b & \text{if } b \leq c \\ \left(1 - \frac{b}{a}\right)^c & \text{if } c \leq b \end{cases}$$

$$= \left[1 - \frac{\max(b,c)}{a}\right]^{\min(b,c)} \tag{17}$$

Hence, the globally feasible approximation (involving selection with replacement) of (7) is:

$$\Phi(A_*) \simeq N \left[1 - \left(1 - \frac{\max(A_*,k)}{N_*}\right)^{\min(A_*,k)}\right] \tag{18}$$

Both alternative expressions making up the globally feasible approximation can be easily inverted for A_* giving rise to a globally feasible estimate for $\Phi^{-1}(A)$ when $A < N$ defined as follows:

$$A_* \simeq \Phi^{-1}(A) = \begin{cases} \dfrac{\log(1 - A/N)}{\log(1 - 1/N)} & \text{if this value is } \leq k \\ N_* \left[1 - (1 - A/N)^{1/k}\right] & \text{if this value is } \geq k \end{cases} \tag{19}$$

Using different likely combinations of N and k parameters, we observed that, when A is not close to N, both branches of (19) actually provide quite similar values which are both very good approximations of A_*. The real difference between them can only be appreciated when A reaches N, as the second branch shows a realistic behavior tending to the value N_*, whereas the first branch would go to infinity. In practice, the second branch is needed only for computing $\overline{A}_*$, yielding $\overline{A}_* \simeq N_*[1 - (2N)^{-1/k}]$. In all the other cases, the first branch can be safely used.

Notice that the approximation (1) proposed by Swamidass and Baldi for $\Phi(A_*)$ coincides with the expression valid for large A_* values in (18) and, thus, the approximation (2) for $\Phi^{-1}(A)$ coincides with the second branch of (19), which corresponds to the viewpoint of a dual experiment. However, in the application practice of compressed chemical fingerprints, owing to fingerprint sparsity, the most appropriate viewpoint would be the primal experiment indeed (e.g., in the study of the ChemDB database [12], A_* has been shown to be always smaller than $k/2$ with a bell-shaped distribution concentrated around a mean value which is smaller than $k/4$). Therefore, a sort of second fortunate choice in [13] has been the introduction of the asymptotic approximation (3) (and its inversion formula (4)), which can be shown to be compatible with the primal experiment viewpoint. In fact, considering the Taylor's expansion of the logarithm, we can write $\log(1 - 1/N) = -1/N - O(1/N^2) \approx -1/N$ when N is large. Hence, also the first branch of (19), valid for the primal experiment, can be asymptotically approximated by $-N \log(1 - A/N)$, that is (4). In any case, the first branch of (19) can be used, as it is, as a globally valid approximation of $\Phi^{-1}(A)$ when $A < N$, approximation which is, in theory, better than (4) when N is not large.

3.3 Experimental Evaluation

Figures 1 and 2 show the predictive power of the various estimates matched against real data, obtained as average values over the generation of 10,000 random fingerprints for each value of the uncompressed weight A_*, ranging from 1 to $A_*^{\max}$ (ten values of $A_*^{\max}$, chosen between 4,000 and 262,144, have been considered).

The experimental setting we used is basically the same as in [13], which correspond to a realistic fingerprint database environment with $N = 512$ and $k = 512$ and, thus, a possible maximum uncompressed fingerprint length $N_* = 262,144$ (in order to test the possible effect of the coincidence of the N and k values, keeping $N_* = 262,144$, we also tried the combination $N = 1,024$ and $k = 256$, for which we obtained perfectly equivalent results).

The graph in Fig. 1 shows, with A ranging from 1 to $N - 1$, four different curves involving the dependence of A_* on A: (i) experimental A_* data, (ii) theoretical $\Phi^{-1}(A)$ data obtained from (7), (iii) estimates of $\Phi^{-1}(A)$ computed via the first branch of (19) and (iv) via (4). The four curves are apparently indistinguishable at the scale of the figure and the relative error of the various estimates with respect to real data is always within a few percentage point (and less than 1% for all $A < 505$). In particular, no significant difference between the prediction accuracy of (4) and either branches of (19) could be evidenced in our simulations for $A < N$. The experimental data plotted in the graph have been generated with $A_*^{\max} = 10,000$ but we registered no appreciable difference with respect to using real data generated by varying the $A_*^{\max}$ value.

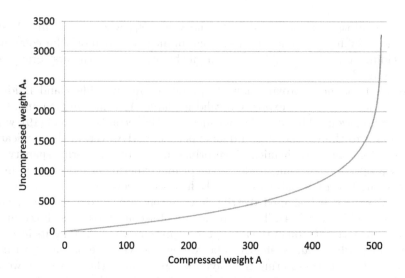

Fig. 1. Experimental evaluation of various A_* estimates for $A < N$ (see text).

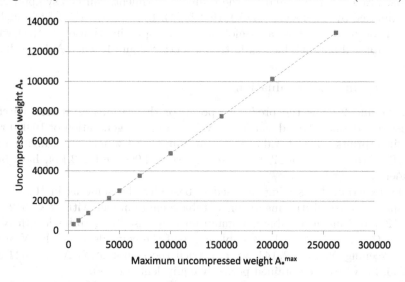

Fig. 2. Experimental evaluation of the A_* estimate when $A = N$.

The graph in Fig. 2 shows what happens to A_* when $A = N$ as a consequence of different values of $A_*^{\max}$ via two series of points drawn for various values of $A_*^{\max}$: (i) experimental A_* data and (ii) estimates provided by the second branch of (16). Also in this case, the two series cannot be distinguished at the scale of the figure and the prediction error of our estimate with respect to real data is always within a few percentage points (and less than 1% for all $A_*^{\max} \geq 10,000$). The linearity of the dependence, as predicted by (16), is evident.

Notice that the only part of the figures which is of interest for practical chemical fingerprint applications is the left portion of the graph in Fig. 1, say with $A < 300$ (in [12], it is shown that A, evaluated over 50,000-molecule samples extracted from the ChemDB database, follows a Gaussian distribution with a mean of about 120 and standard deviation of about 40).

4 Conclusions

In this paper, we reconsidered the theoretical background underlying the approach in [13], based on the use of formula (4) for improving the similarity measure between compressed chemical fingerprints.

First, we applied the γ-transform approach and some standard techniques to the probabilistic characterization of the fingerprint compression process. In particular, we showed that the real value of $E[A_*|A]$ is quite different from the approximation determined in [13].

Second, we observed that, taking into account the sparsity and the generation process of the fingerprints, the use of $E[A_*|A]$ is inappropriate for estimating A_* from A and $\Phi^{-1}(A)$ has to be used instead. Moreover, in the light of the globally feasible approximation concept, we tried to improve the formula (4) used to this end in [13].

Furthermore, several experiments with random generated fingerprints have been executed and their outcomes basically demonstrated the correctness of the proposed analysis and the goodness of the resulting estimates (16) and (19). However, experiments also confirmed the validity of the estimate (4) for practical application purposes.

In conclusion, we ended up in confirming the validity of the Swamidass and Baldi correction approach presented in [13], based on the use of formula (4) for improving the similarity measure between compressed fingerprints. In particular, we provided for it a novel and more solid theoretical underpinning, which complements the original experimental validation with actual data given in [13].

References

1. Aung, Z., Ng, S.-K.: An indexing scheme for fast and accurate chemical fingerprint database searching. In: Gertz, M., Ludäscher, B. (eds.) SSDBM 2010. LNCS, vol. 6187, pp. 288–305. Springer, Heidelberg (2010). https://doi.org/10.1007/978-3-642-13818-8_22
2. Bohacek, R.S., McMartin, C., Guida, W.C.: The art and practice of structure-based drug design: a molecular modeling perspective. Med. Res. Rev. **16**(1), 3–50 (1996)
3. Daylight Chemical Information Systems Inc. http://www.daylight.com/
4. Flower, D.R.: On the properties of bit string-based measures of chemical similarity. J. Chem. Inform. Comput. Sci. **38**(3), 378–386 (1998)
5. Grandi, F.: Advanced access cost models for databases. Ph.D. Dissertation (in Italian), DEIS, University of Bologna, Italy (1994)

6. Grandi, F.: On the signature weight in "multiple" m signature files. ACM SIGIR Forum **29**(1), 20–25 (1995)
7. Grandi, F.: The γ-transform approach: a new method for the study of a discrete and finite random variable. Int. J. Math. Models Methods Appl. Sci. **9**, 624–635 (2015). http://www.naun.org/main/NAUN/ijmmas/2015/b442001-411.pdf
8. Grandi, F.: On the analysis of Bloom filters. Inf. Process. Lett. **129**, 35–39 (2018)
9. Grandi, F., Scalas, M.R.: Block access estimation for clustered data using a finite LRU buffer. IEEE Trans. Softw. Eng. **19**(7), 641–660 (1993)
10. Hajiebrahimi, A., Ghasemi, Y., Sakhteman, A.: FLIP: an assisting software in structure based drug design using fingerprint of protein-ligand interaction profiles. J. Mol. Graph. Model. **78**, 234–244 (2017)
11. Rouvray, D.: Definition and role of similarity concepts in the chemical and physical sciences. J. Chem. Inf. Comput. Sci. **32**(6), 580–586 (1992)
12. Swamidass, S.J., Baldi, P.: Statistical distribution of chemical fingerprints. In: Bloch, I., Petrosino, A., Tettamanzi, A.G.B. (eds.) WILF 2005. LNCS (LNAI), vol. 3849, pp. 11–18. Springer, Heidelberg (2006). https://doi.org/10.1007/11676935_2
13. Swamidass, S.J., Baldi, P.: Mathematical correction for fingerprint similarity measures to improve chemical retrieval. J. Chem. Inf. Model. **47**(3), 952–964 (2007)
14. Tversky, A.: Features of similarity. Psychol. Rev. **84**(4), 327–352 (1977)
15. Yao, S.B.: Approximating block accesses in database organizations. Commun. ACM **20**(4), 260–261 (1977)

Time Series Retrieval Using DTW-Preserving Shapelets

Ricardo Carlini Sperandio[1]([⊠]), Simon Malinowski[2]([⊠]), Laurent Amsaleg[3]([⊠]), and Romain Tavenard[4]([⊠])

[1] IRISA-Inria, Rennes, France
ricardo.carlini-sperandio@irisa.fr
[2] IRISA-Univ. Rennes 1, Rennes, France
simon.malinowski@irisa.fr
[3] CNRS-IRISA, Rennes, France
Laurent.Amsaleg@irisa.fr
[4] Univ. Rennes 2, Rennes, France
romain.tavenard@univ-rennes2.fr

Abstract. Dynamic Time Warping (DTW) is a very popular similarity measure used for time series classification, retrieval or clustering. DTW is, however, a costly measure, and its application on numerous and/or very long time series is difficult in practice. This paper proposes a new approach for time series retrieval: time series are embedded into another space where the search procedure is less computationally demanding, while still accurate. This approach is based on transforming time series into high-dimensional vectors using DTW-preserving shapelets. That transform is such that the relative distance between the vectors in the Euclidean transformed space well reflects the corresponding DTW measurements in the original space. We also propose strategies for selecting a subset of shapelets in the transformed space, resulting in a trade-off between the complexity of the transformation and the accuracy of the retrieval. Experimental results using the well known UCR time series demonstrate the importance of this trade-off.

1 Introduction

Time series data is massively produced 24 * 7 by millions of users, worldwide, in domains such as finance, agronomy, health, earth monitoring, weather forecasting, multimedia, etc. Due to the advances in sensor technology and its proliferation, applications may produce millions to trillions of time series per day, making time series data-mining further challenging.

Most time series data-mining algorithms rely on the Dynamic Time Warping (DTW) measure at their core, which proves to return high-quality results, but which is also very costly to compute [3]. Many researchers have attempted to reduce its cost in order to run it at scale. Numerous smart optimizations, diverse lower bounds and other techniques have been applied [4], however, the quest for processing extremely large collections of time series is still active.

ⓒ Springer Nature Switzerland AG 2018
S. Marchand-Maillet et al. (Eds.): SISAP 2018, LNCS 11223, pp. 257–270, 2018.
https://doi.org/10.1007/978-3-030-02224-2_20

This paper focuses on the task of time series retrieval according to the DTW measure. The classical scenario for this task is the following: let τ be a test time series (query), retrieval is about finding a time series in a dataset $\mathcal{T}$ which is the closest one to τ with respect to the DTW measure. In other words, it is finding T^*, such that:

$$T^* = \arg \min_{T_i \in \mathcal{T}} DTW(\tau, T_i) \tag{1}$$

A traditional way to identify T^* relies on the brute force DTW computation between τ and all series of $\mathcal{T}$. This approach is not tractable when dealing with long time series or huge data sets. Hence, approximated methods are preferred, aiming at reducing the retrieval cost while being as accurate as possible.

We propose here an approximate time series retrieval approach based on the shapelet transform. The basic idea of the proposed approach is to transform the time series of the dataset into vectorial representations such that a Euclidean search can then be efficiently applied to retrieve the nearest neighbour of the transformed query. Of course, the transformation needs to be carefully designed so that the approximate search is accurate enough. Another crucial point is related to the computing cost of the transformation. At test time, the query needs to be first transformed before being compared with transformed time series of the dataset. Hence, the transformation should not be too costly.

In this paper, DTW-preserving shapelets [11] are used to transform time series into vectorial representations. This transform is such that the relative Euclidean distance in the transformed space well reflects the original DTW measurements. They are hence well adapted to our task. Transforming time series has a cost, but it can be traded-off against the accuracy of the retrieval by selecting a varying number of shapelets for computing the transformation. To tackle this issue, three shapelet selection strategies are proposed in this paper. These strategies create a trade-off between complexity and accuracy of the retrieval task.

The remainder of this paper is organized as follows: Sect. 2 presents related work. The proposed Learning DTW-Preserving Shapelet Retrieval (DPSR) method is presented in Sect. 3. Section 4 presents experimental results and conclusions and future work are presented in Sect. 5.

2   Related Work

Various similarity measures have been used for time series retrieval, and excellent surveys provide a good coverage of their pros and cons [4,20]. In a nutshell, the straightforward Euclidean distance is not robust to distortions and time shifts, and is hence not very appropriate for time series. Approaches based on the DTW are much preferred because it is able to find the optimal alignment between two given time series, thus coping with local distortions along the time-line. DTW, however, is costly to calculate because of its quadratic time complexity.

Several approaches have therefore been proposed in order to speed up the computation of DTW, including restricting around the diagonal of the distance matrix [7,16] or the design of lower bounds. The most popular lower bound is the

LB_Keogh lower bound (LB_Keogh) [8]. It first needs to rescale each time series to the same length and then it builds their envelopes by accounting for their maximum and minimum values inside a sliding window. The distance between two time series returned by LB_Keogh corresponds to the areas of their envelopes that do not overlap. In [14], Rakthanmanon et al. proposed the UCR Suite framework which includes several acceleration approaches that can be combined in order to index time series under the DTW measure. Recently, Tan et al. in [18] proposed an adaptation of Priority Search K-means to index time series embedded in a space induced by DTW. Interested readers should read three comprehensive reviews [3,4,13].

The Piecewise Aggregate Approximation (PAA) [9,22] has been used to build iSAX [17], one of the most famous time series indexing system. PAA is a transformation that divides time series into smaller pieces and create an approximation of each piece. This transformation is at the core of iSAX, which was shown to be very accurate and fast for retrieving time series. However, it is based on the Euclidean distance and not the DTW.

We propose a retrieval scheme based on a transformation preserving the DTW. This transformation makes use of shapelets. Shapelets were introduced by Ye and Keogh in [21] for time series classification. The underlying intuition behind shapelets is that time series belonging to one class are likely to share some common subsequences. Shapelets were therefore originally defined as existing subsequences of time series that best discriminate classes.

Hills et al. then build on the idea of shapelets by proposing the Shapelet Transform [6]. In this approach, each time series is transformed into a vector whose components represent the distances between the time series and the shapelets. Transforming a time series into its vectorial representation requires to slide each shapelet against that time series in order to find the best matching locations and then compute the corresponding distances. The cost of shapelet transform is therefore highly dependant on the number of shapelets that are used to create vectorial representations.

Instead of using existing subsequences as shapelets, Grabocka et al. in [5] propose to rather forge the shapelets by learning the subsequences that minimize a classification loss. Shapelets have also been used for unsupervised tasks and not only for classification. In [11], Lods et al. learn shapelets such that the resulting vectorial representation preserves as well as possible the DTW distance between raw time series, targeting time series clustering. In [23], Zakaria et al. extract the shapelets dividing the set of time series into well separated groups.

3   Time Series Retrieval with DTW-Preserving Shapelets

In this section, we detail our approach for time series retrieval under DTW. This approach builds on (i) the shapelet transform paradigm and (ii) the DTW-preserving shapelets that are proposed in [11]. We first quickly review these building blocks before presenting the design of our retrieval scheme.

3.1   Background on Shapelets and Shapelet Transform

A shapelet $S = s_1, \ldots, s_l$ is a temporal sequence (that can be extracted from existing time series, or learned). The distance between S and a time series $T = t_1, \ldots, t_L$ is defined as:

$$d(T, S) = \min_{1 \leq j \leq L-l+1} \sqrt{\sum_{i=1}^{l} (s_i - t_{i+j-1})^2} \qquad (2)$$

In other words, Euclidean distances between S and every subsequence of T (of length l) are computed and only the best match (minimum distance) is kept. The shapelet transform of a time series was proposed in [6] for time series classification. It is a two step process: (i) selecting an appropriate set of shapelets $S = \{S_1, \ldots, S_K\}$ and (ii) transforming time series into Euclidean vectors. During the second step, each time series T is transformed into a vector $v_1, \ldots, v_K$ such that $v_i = d(T, S_i)$, $1 \leq i \leq K$, where $S_1, \ldots, S_K$ are the shapelets that were selected during the first step. The dimensions of v represent the distance between T and the shapelets of S. Such representations of time series are then used to feed a classifier, when the targeted application is classification.

3.2   Learning DTW-Preserving Shapelets

In [11], Lods *et al.* propose to learn a set of shapelets such that the shapelet-transformed representation preserves as well as possible the original DTW measure. The shapelets are learned such that Euclidean distance in the transformed space approximates the DTW. Such shapelets can hence be used for unsupervised tasks, that is, where no labels are available.

The approach proposed by Lods *et al.* relies on minimizing the loss:

$$\mathcal{L}(T_i, T_j) = \frac{1}{2} \left(\text{DTW}(T_i, T_j) - \beta \left\| \overline{T}_i - \overline{T}_j \right\|_2 \right)^2, \qquad (3)$$

where β is a scale parameter, learned and $\overline{T}_i$ represents the shapelet transform of T_i. The overall loss for a dataset $\mathcal{T}$ of N time series is given by:

$$\mathcal{L}(\mathcal{T}) = \frac{2}{N(N-1)} \sum_{i=1}^{N} \sum_{j=i+1}^{N-1} \mathcal{L}(T_i, T_j). \qquad (4)$$

The minimization of this loss is done via a stochastic gradient descent with respect to β and S. Once the shapelets and the parameter β are learned, they can be used to transform every time series into a Euclidean vector.

3.3   Transforming Times Series for Retrieval Under DTW

In this paper, we propose a retrieval scheme based on transforming time series into high-dimensional vectors. Its offline step transforms all time series of the data set into their vectorial representation, as explained above. Its online step uses a query to probe the dataset. To do so, the vectorial representation of the query is first determined. Then, Euclidean distances to the transformed time series of the dataset are computed. This results in a list of transformed time series that are ranked according to their proximity to the transformed query. The nature of the transformation that is using DTW-preserving shapelets is such that this ranking in the transformed space is an approximation of the ranking that would be produced in the original space according to the DTW measure. However, this approximate ranking is obtained much faster, as Euclidean measurements are cheaper to obtain compared to DTW measurements.

Two ways to process that ranked list of time series in the transformed space can be designed: (a) the original time series associated to the top-ranked in this list is considered to be the nearest neighbour of the query, or (b) the true DTW is computed between the original untransformed query and the original version of the first few elements of that list in order to refine the search. (a) puts a lot of pressure on the quality of the shapelet transform because the nearest time series under DTW has also to be the closest in the Euclidean space, and this for any time series in the dataset and any query. Odds of degrading quality in comparison to what the true DTW would determine are high, but this method is extremely fast. In contrast, (b) is more demanding because more DTW are computed, but it is also returns better quality results as more time series are scrutinized. In this case, it matters that the closest time series under DTW belongs to these first few elements, instead of being ranked first.

Overall, two properties are important for an approximate retrieval scheme: (i) the complexity of the transformation should be small to reduce the overhead induced by transforming the query and (ii) the true nearest neighbour has to be as close as possible to the first element of the list of approximate nearest neighbours. In other words, the transformation should preserve the ranking.

The retrieval scheme proposed in this paper focuses on these two important properties.

3.4   Ranking and Selecting Shapelets

The transformation of the query before searching in the transformed space is costly. Indeed, the computation of one coordinate of the transformed vector requires sliding a shapelet over the query τ and finding the best match. It is particularly costly for long time series and when the number of learned shapelets is high. We hence focus in this section on ranking and selecting a subset of shapelets most suited to the retrieval task.

General considerations about high-dimensional representations suggest that components of vectors might not all be equally useful. This well known observation led to designing dimensionality reduction methods performing feature selection. Feature selection algorithms typically use (i) an evaluation metric to compare different feature subsets, (ii) a strategy for building consecutive subsets and (iii) a stopping criterion [1,10,12,15]. The characteristics that we use to design our shapelet selection algorithm are described hereafter. Selecting a few appropriate shapelets saves computations at transform time without degrading significantly the quality of the approximation. Algorithm 1 is the corresponding pseudo-code.

Evaluation Metric to Compare Shapelet Subsets. To compare the performance of different shapelet subsets, we need a groundtruth based on the true DTW between time series. To build that groundtruth, the DTW between all time series pairs in the training set is computed and we record for each time series the identifier of its nearest-neighbour. This is done once only, off-line.

We use a 10-fold validation setup in order to evaluate the performance of a subset. For one transformed time series in the validation set (query), we rank all the transformed time series in the training set according to their Euclidean distance to the query. It is therefore possible to determine at which rank the true 1-nearest neighbor time series is. Repeating this operation for all the validation time series and for all the folds amounts to building an histogram of the rank at which the true nearest-neighbour appear. This histogram can also be interpreted as the empirical probability of observing the true 1-nearest neighbor time series at any particular rank after the transform. This histogram can therefore be considered to be a probability density function (after a proper normalization though). From this PDF, it is straightforward to construct its natural counterpart which is the CDF, the cumulative distribution function, and to compute the associated area under the curve (AUC). We consider this AUC value as the performance measure to evaluate the quality of a shapelet subset. The higher that AUC, the better the shapelet subset. This metric is well adapted to the task of nearest-neighbour retrieval as it favors high ranking of the true nearest neighbor in the approximated list.

Shapelet Subset Selection. To select the best subset of shapelets, an exhaustive selection method can be applied. However, in this case, the computational cost is prohibitively high. We have chosen a greedy-based forward selection method, that is classically used in the feature selection domain.

Our procedure to select the best shapelet subset begins with an empty list. Then it iteratively adds shapelets that best improve the quality of the subset (by measuring the resulting AUC), one by one, until a stopping criterion is met (the different stopping criterion we used are described in the following).

Stopping Criterion. We define three different stopping criterion, that determine when to stop adding shapelets to the current set of selected shapelets:

1. Global maximum ($DPSR_g$): Shapelets are added one by one until no more shapelets are available. At the end, the subset that leads to the best overall AUC is selected.
2. Tangent ($DPSR_t$): We compute the normalized slope between the AUC value of the current selected subset and the one obtained by adding the shapelet that best improves the AUC. If this slope is less than 1, then the shapelet selection is stopped.

Algorithm 1. Shapelet ranking and selection

1: **input:** $\mathcal{S}$ {Set of learned shapelets}, $\mathcal{T}$ {Time series train set}
2: **output:** $\mathcal{S}_s$ {Ranked list of selected shapelets}
3: $\mathcal{S}_a \leftarrow \mathcal{S}$ {Shapelets available to evaluate, initially all}
4: $\mathcal{S}_s \leftarrow \emptyset$ {Shapelets selected, initially empty}
5: $stop \leftarrow$ FALSE
6: **repeat**
7:    $score_b \leftarrow -1$
8:    **for all** $S \in \mathcal{S}_a$ **do**
9:       $\mathcal{S}_t \leftarrow \mathcal{S}_s \cup S$, $score \leftarrow$ AUC value of the set $\mathcal{S}_t$
10:      **if** $score > score_b$ **then** $score_b \leftarrow score$, $S_b \leftarrow S$
11:   **end for**
12:   $\mathcal{S}_s \leftarrow \mathcal{S}_s \cup S_b$, $\mathcal{S}_a \leftarrow \mathcal{S}_a \setminus S_b$
13:   **if** Stopping criterion is met **then** $stop \leftarrow$ TRUE
14: **until** $stop =$ TRUE

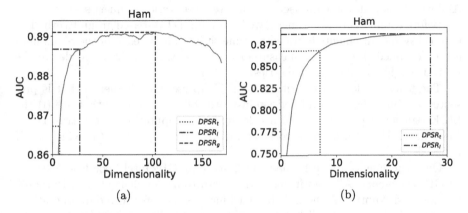

Fig. 1. (a) Impact of the three different stopping criterion in terms of number of selected shapelets for the Ham dataset. (b) Zoom on the first dimensions of (a).

3. Local maximum ($DPSR_l$): The shapelet selection is stopped as soon as adding a shapelet does not improve the AUC value.

Figure 1 shows the AUC values at each iteration of the shapelet selection algorithm on the Ham dataset (from the UCR-UEA archive [2]). For this dataset, it has been learned 170 shapelets using Lods *et al.* approach ($DPSR_f$). We can see on this figure the impact of the 3 different criteria. If the tangent criterion ($DPSR_t$) is used, then the selection process would end with 7 shapelets (Fig. 1b, which zooms on the first 30 dimensions), while 27 would be selected by the local maximum criterion ($DPSR_l$) and 103 by the global maximum one ($DPSR_g$). Note that for the global maximum criterion the selection process cannot be stopped before having ranked and selected all shapelets one by one, contrary to the two other criteria for which the process can be stopped as soon as the criterion is reached. We illustrate in the experiments the trade-off between dimensionality and accuracy induced by these criteria.

4   Experiments

This section presents an experimental evaluation of our DPSR approach. The trade-off between accuracy and computation complexity is discussed. Performance of DPSR is compared to Piecewise Aggregate Approximation (PAA) [9,22] and LB_Keogh lower bound (LB_Keogh) [8]. We start with our experimental setup.

4.1   Experimental Setup

We consider the 85 datasets from the well known UCR-UEA Time Series Archive [2]. For all these datasets, we use test sets series as queries.

Before running any experiment, we built a full DTW-based groundtruth. The true DTW measurements between all time series pairs in each of the 85 families are determined. From these measurements, it is straightforward to derive the nearest-neighbour of each time series.

The feature selection algorithm for DTW-preserving Shapelets that is presented in this paper is evaluated against two solid competitors that are (i) the LB_Keogh lower-bound used to accelerate the computation of the true DTW (used in the UCR Suite framework) and (ii) the PAA approach (on which iSAX is based).

Our experiments are performed on a 24-core 2.8 GHz Intel Xeon ES-2630 with 64 GB of memory. All algorithms and structures are implemented in Python3, Cython and NumPy. Although the machine has 24 cores, the only operations using parallelism are the distance computations handled by NumPy during the course of each experiment—no other parallelism is enforced. The tslearn [19] toolkit was used for the computation of PAA and LB_Keogh.

For each family in the UCR archive, a set of DTW-preserving shapelets is learned using the algorithm proposed in [11], with default parameters. To learn high-quality shapelets, 500,000 iterations of the gradient descent algorithm are performed. In addition, two sets of shapelets are learned for each family of time series, and the one with the smallest overall loss is selected.

4.2   Dimensionality Versus Accuracy: Reaching a Plateau

This first experiment aims at comparing the performance of PAA and DPSR for time series retrieval. This comparison is performed in terms of trade-off between AUC and the dimensionality of the transformation. Transformations range from very rough approximations (few shapelets, few pieces for PAA) to finer grain representations. For consistency, we compared PAA and DPSR for the same dimensionality. Figure 2 illustrates the resulting AUC values for two specific data sets, Gun_Point and Beef. With Gun_Point, DPSR outperforms PAA for

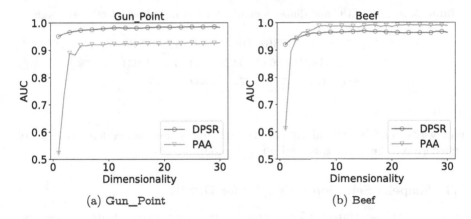

Fig. 2. Comparing DPSR and PAA for two specific time series. The ability to retrieve the correct time series from the groundtruth is represented by the resulting AUC.

Table 1. Comparing DPSR and PAA for the full UCR Archive at different dimensionalities. Number of times each method outperforms the other in terms of AUC is reported together with the average AUC values over the 85 datasets.

Dimensionality	5	10	20	30
# DPSR wins	70	69	73	72
# PAA wins	15	16	12	13
Avg. AUC for PAA	0.838	0.847	0.844	0.841
Avg. AUC for DPSR	0.906	0.921	0.929	0.932

all dimensionalities. The results for `Beef` are more contrasted: when the time series is split into more than 5 pieces, PAA outperforms the shapelet based approach. Please note that these figures show quality measures for the first 30 dimensions only. Considering more than 30 dimensions does not provide any useful extra information for these datasets.

Table 1 compiles the comparison of PAA and the approach based on DPSR for all 85 time series by counting the number of times each method performs better than the other, for dimensionalities 5, 10, 20 and 30. We also report the average AUC values at these dimensionalities for the two different approaches. Overall, this table shows that DPSR consistently outperforms PAA for all dimensionalities. We can also observe that both methods seem to reach an AUC plateau, sooner for PAA than it is for DPSR.

For DPSR, this plateau highlights the importance of selecting shapelets. It indicates that the original approach by Lods *et al.* creates far too many shapelets, and that a lot of them do not contribute significantly to enhancing the resulting vectorial representation. Generating so many shapelets is wasting computing resources at transform time because many shapelets have to be slid, some in

Table 2. Average AUC and dimensionalities for feature subset selection strategies.

	DPSR$_t$	DPSR$_l$	DPSR$_g$	DPSR$_f$
Avg. AUC	0.906	0.928	0.935	0.934
Avg. Dim.	4.4	27.3	54.0	156.1

pure waste. This is particularly important at query time, as reducing the cost of transforming the query time series is paramount.

4.3 Shapelet Selection Strategies for DPSR

In Sect. 3.4, three strategies for stopping aggregating selected features were presented. We now evaluate their effectiveness, which is a trade-off between their accuracy in terms of AUC and the transformation cost they cause. Typically, small subsets allow for very fast transforms (just a few shapelets need to be slid at test time) but quality is typically low, whereas in contrast larger subsets improves AUC performance but cause more expensive transform operations.

To observe this trade-off, we selected by cross-validation on the training sets of each dataset the best shapelet subset for the three different stopping criterion. We then used these subsets at test time for approximate retrieval. The AUC performance of the three stopping strategies has been computed on the 85 considered datasets. The average AUC value is given in Table 2, together with the average number of selected shapelets. The last column of Table 2 (DPSR$_f$) corresponds to the case where no shapelet selection is performed (i. e., the whole shapelet set learned beforehand is used).

We can observe that the three proposed criteria generate a trade off between accuracy of the retrieval (AUC) and computational time of the transform (linear with the number of shapelets). DPSR$_t$, the most aggressive strategy, selects very few shapelets (a little bit more than 4, on average) for an average AUC of 0.906. DPSR$_g$, the most conservative strategy uses on average 54 shapelets, but the corresponding quality improvement is quite small: it goes from 0.906 to 0.935. This is a clear illustration of the trade off, also exemplified by the DPSR$_l$ strategy, which is in between these two strategies.

We can also observe the importance of the feature selection itself: when no selection is made (DPSR$_f$), the average AUC is slightly lower than for DPSR$_g$ and the corresponding average number of shapelets used is almost tripled.

4.4 Feature Selection Versus Constrained Feature Learning

The previous experiment demonstrated that only a small fraction of the learned shapelets are truly useful because they significantly contribute to the quality of the retrieval. We show here that it is the combination of the learning stage and the feature selection strategy that leads to such behaviour. For that purpose, we compare the AUC performance of the proposed retrieval scheme (DPSR$_t$ and

DPSR$_l$) with a method where the same number of shapelets is directly learned using the algorithm presented in [11].

We used the previous experiment to record for each dataset how many shapelets the DPSR$_t$ and DPSR$_l$ selected. Then, we ran the shapelet learning algorithm of [11] using that number of shapelets (for each dataset and for each DPSR strategy). We know from the previous experiments that the average AUC for DPSR$_t$ over all the datasets is 0.906, with using 4.4 shapelets on average. When directly learning that same number of shapelets, then the average AUC over all datasets is 0.866. Furthermore, considering individually the 85 families of time series, the DPSR$_t$ strategy performs better than the one directly learning the appropriate number of shapelets in 74 cases (out of 85). Same conclusions can be drawn with DPSR$_l$. The average AUC of DPSR$_l$ is 0.928, whereas the AUC obtained when directly learning the same number of shapelets is equal to 0.905. The DPSR$_l$ strategy wins 71 times out of 85 in that case.

These results indicate that it is worth spending more time offline to learn a huge set of shapelets and then selecting the more appropriate. This is better than trying to save computational time by learning less shapelets. Also, our feature selection strategy allows to decide on the number of shapelets in a data driven fashion, contrary to a purely heuristic approach.

4.5   Comparing Methods at Their Best

In this experiment, we compare the respective performance of DPSR, PAA and LB_Keogh when their parameters (number of segments for PAA, subset of shapelets for DPSR and window length for LB_Keogh) are cross-validated on the train set. The results comparing the performance of (i) DPSR and PAA methods and (ii) DPSR and LB_Keogh are given in the Table 3. This table gives the number of times each method wins over the other. Overall, DPSR outperforms PAA even for the DPSR$_t$ criterion. Interestingly, comparing the dimensionalities when DPSR or PAA are winning provides insightful results. Consider for example the DPSR$_t$ strategy. That strategy wins over PAA 54 times. Among these 54 wins, in 45 cases, DPSR needs fewer dimensions than PAA. It means that in 9 cases, DPSR needs more dimensions than PAA to provide better results. The dual point of view is also insightful: PAA outperforms the DPSR$_t$ method in 31 cases, but all PAA transformations need more pieces than DPSR$_t$. Not only DPSR wins more frequently than PAA, but when it wins, it is with shorter representations. This is also true for DPSR$_l$ and DPSR$_g$. The average AUC value for PAA is equal to 0.866 which is worse than the average AUC value of the 3 DPSR approaches. Against LB_Keogh, DPSR is only worse for the DPSR$_t$ criterion which leads to a very small representation. The average value of AUC for LB_Keogh is 0.908. It is better than DPSR$_t$, but not than DPSR$_l$ and DPSR$_g$.

An important observation is that, unlike LB_Keogh and PAA, DPSR allows comparison between time series of different lengths.

Table 3. Comparing DPSR, PAA and LB_Keogh with their best parameters (learned by cross-validation). We report here the number of times each method outperforms the other in terms of AUC.

	DPSR vs. PAA		DPSR vs. LB_Keogh	
	# wins for DPSR	# wins for PAA	# wins for DPSR	# wins for LB_Keogh
DPSR$_t$	54	31	33	52
DPSR$_l$	63	22	62	23
DPSR$_g$	70	15	64	21

4.6 Search Costs

So far, only quality comparisons have been done. We observe now the computational costs of the approaches discussed here. Additional experiments were performed for recording search times when the dimensionality of the representations for DPSR and PAA gradually increase. Search times for Gun_Point and Beef are plot on Fig. 3 (the plots show only results for their first 60 dimensions), and the average search times for the full UCR archive is on Fig. 3c.

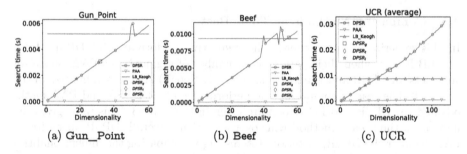

(a) Gun_Point   (b) Beef   (c) UCR

Fig. 3. Average Search Times for two specific time series and for the full UCR Archive. Varying dimensionality.

These figures also show the time it takes to compute only the envelopes of time series for LB_Keogh. That process, coupling LB_Keogh and DTW is guaranteed to find the same time series as the one indicated in the groundtruth. The quality of any approximate search scheme can only be equal or lower. But the time for solely computing the LB_Keogh value on all time series is the absolute minimal cost the real LB_Keogh+DTW could have.

These figures show that the time for computing envelopes with LB_Keogh is fixed, which is normal. They also show that PAA is the fastest approach. Its underlying principles are simple and cause light computations, the search times increasing slightly with the dimensionality. The time taken per search for DPSR is also increasing with the number of dimensions: there are more and more shapelets to slide, and distance computations are more demanding. Three remarkable signs are placed on the search time plot for the DPSR approach. They refer to the search times observed when the number of shapelets in use correspond to what $DPSR_t$, $DPSR_l$ and $DPSR_g$ determined.

5   Conclusions and Future Work

In this work, we have presented an approach for time series retrieval based on learning DTW-preserving shapelets (DPSR). This approach first transforms time series into high dimensional vectors such that the Euclidean distance between the vectors in the transformed space well reflects the DTW measurements in the original space. This targets preserving the quality of time series retrieval compared to the DTW. Relying on Euclidean distances is more efficient than computing the costly DTW measures. This targets computational efficiency, facilitating time series retrieval at scale.

In order to cope with larger scales, we also propose different shapelet selection strategies to trade complexity of the retrieval against accuracy. Even the most aggressive strategy (that selects very few shapelets) provides reasonable accuracy. Experimental results show the importance of this feature selection.

This work is a first step into the design of a time series indexing system. At very large scale, the many high dimensional vectors representing time series could be inserted into an index, avoiding the exhaustive Euclidean distance calculations, further improving performance. Such an approach can be advantageously used for *anytime indexing* of time series.

Acknowledgments. The current work has been performed with the support of CNPq (*Conselho Nacional de Desenvolvimento Científico e Tecnológico*), Brazil (Process number 233209/2014–0). The authors are grateful to the TRANSFORM project funded by STIC-AMSUD (18-STIC-09) for the partial financial support to this work.

References

1. Blum, A., Langley, P.: Selection of relevant features and examples in machine learning. Artif. Intell. **97**(1–2), 245–271 (1997)
2. Chen, Y., et al.: The UCR time series classification archive, July 2015. www.cs.ucr.edu/~eamonn/time_series_data/
3. Ding, H., Trajcevski, G., Scheuermann, P., Wang, X., Keogh, E.J.: Querying and mining of time series data: experimental comparison of representations and distance measures. PVLDB **1**(2), 1542–1552 (2008)
4. Esling, P., Agón, C.: Time-series data mining. CSUR **45**(1), 12:1–12:34 (2012)
5. Grabocka, J., Schilling, N., Wistuba, M., Schmidt-Thieme, L.: Learning time-series shapelets. In: KDD, pp. 392–401. ACM (2014)
6. Hills, J., Lines, J., Baranauskas, E., Mapp, J., Bagnall, A.: Classification of time series by shapelet transformation. DMKD **28**(4), 851–881 (2014)
7. Itakura, F.: Minimum prediction residual principle applied to speech recognition. IEEE Trans. Sig. Process. **23**(1), 67–72 (1975)
8. Keogh, E.J.: Exact indexing of dynamic time warping. In: VLDB, pp. 406–417. Morgan Kaufmann, Burlington (2002)

9. Keogh, E.J., Chakrabarti, K., Pazzani, M.J., Mehrotra, S.: Dimensionality reduction for fast similarity search in large time series databases. KAIS **3**(3), 263–286 (2001)

10. Li, J., et al.: Feature selection: a data perspective. ACM Comput. Surv. **50**(6), 94:1–94:45 (2017)

11. Lods, A., Malinowski, S., Tavenard, R., Amsaleg, L.: Learning DTW-preserving shapelets. In: Adams, N., Tucker, A., Weston, D. (eds.) IDA 2017. LNCS, vol. 10584, pp. 198–209. Springer, Cham (2017). https://doi.org/10.1007/978-3-319-68765-0_17

12. Moradi, P., Rostami, M.: A graph theoretic approach for unsupervised feature selection. Eng. Appl. AI **44**, 33–45 (2015)

13. Papapetrou, P., Athitsos, V., Potamias, M., Kollios, G., Gunopulos, D.: Embedding-based subsequence matching in time-series databases. TODS **36**(3), 17:1–17:39 (2011)

14. Rakthanmanon, T., et al.: Searching and mining trillions of time series subsequences under DTW. In: KDD, pp. 262–270. ACM (2012)

15. Saeys, Y., Inza, I., Larrañaga, P.: A review of feature selection techniques in bioinformatics. Bioinformatics **23**(19), 2507–2517 (2007)

16. Sakoe, H., Chiba, S.: Dynamic programming algorithm optimization for spoken word recognition. IEEE Trans. Sig. Process. **26**(1), 43–49 (1978)

17. Shieh, J., Keogh, E.J.: iSAX: indexing and mining terabyte sized time series. In: KDD, pp. 623–631. ACM (2008)

18. Tan, C.W., Webb, G.I., Petitjean, F.: Indexing and classifying gigabytes of time series under time warping. In: SDM, pp. 282–290. SIAM (2017)

19. Tavenard, R.: tslearn: a machine learning toolkit dedicated to time-series data (2017). https://github.com/rtavenar/tslearn

20. Wang, X., Mueen, A., Ding, H., Trajcevski, G., Scheuermann, P., Keogh, E.J.: Experimental comparison of representation methods and distance measures for time series data. DMKD **26**(2), 275–309 (2013)

21. Ye, L., Keogh, E.J.: Time series shapelets: a new primitive for data mining. In: KDD, pp. 947–956. ACM (2009)

22. Yi, B., Faloutsos, C.: Fast time sequence indexing for arbitrary Lp norms. In: VLDB, pp. 385–394. Morgan Kaufmann, Burlington (2000)

23. Zakaria, J., Mueen, A., Keogh, E.J.: Clustering time series using unsupervised-shapelets. In: ICDM, pp. 785–794. IEEE Computer Society (2012)

Author Index

Printed in the United States
By Bookmasters

Printed in the United States
By Bookmasters